—PROVEN TECHNIQUES FOR—
KEEPING HEALTHY
CHICKENS

THE BACKYARD GUIDE TO RAISING CHICKS, HANDLING BROODY HENS, BUILDING COOPS, AND MORE

CARISSA BONHAM
FOREWORD BY JANET GARMAN

Skyhorse Publishing

To Super Asher and The Chicken King:
Thanks for being my co-adventurers in chicken keeping.

Skyhorse Publishing books may be purchased in bulk at special discounts for sales promotion, corporate gifts, fund-raising, or educational purposes. Special editions can also be created to specifications. For details, contact the Special Sales Department, Skyhorse Publishing, 307 West 36th Street, 11th Floor, New York, NY 10018 or info@skyhorsepublishing.com.

Skyhorse® and Skyhorse Publishing® are registered trademarks of Skyhorse Publishing, Inc.®, a Delaware corporation.

Visit our website at www.skyhorsepublishing.com.

10 9 8

Library of Congress Cataloging-in-Publication Data is available on file.

Cover design by Mona Lin
Cover photo credit: Carissa Bonham

Print ISBN: 9781510737204
Ebook ISBN: 9781510737228

Printed in China

CONTENTS

Foreword

I am always excited to see a new chicken care book hit the shelves. *Proven Techniques for Keeping Healthy Chickens* meets all the standard requirements of a chicken care book for modern backyard poultry owners but takes things to a higher level.

Ms. Bonham goes beyond the usual food, shelter, and basic how-to instructions. Of course, you need a coop, but did you know you should evaluate where to place the coop? Do you want to know how to add herbs to your chicken coop for aroma therapy and chicken health? That's in this book. How about cooking for your chickens? Yes, they love homemade treats, and the recipes in the book will have your flock eating out of your hand.

Journey through necessary tips that relate to predator safety, coop cleanliness, and which kitchen scraps should go to the feathered family. Then enjoy all the fun boredom-buster ideas to enrich your chicken's daily lives. Maybe you want to watch a fun game of Jump at the Piñata or teach your chicken a new trick.

This book is invaluable for the new flock owner and will add to the knowledge of an experienced chicken keeper. In addition, the photos throughout the book are gorgeous, full-color glimpses into the lives of chickens. *Proven Techniques for Keeping Healthy Chickens* by Carissa Bonham is a valuable new resource for the backyard poultry world.

Janet Garman
owner of Timber Creek Farm (timbercreekfarmer.com)
& author of *50 Do-It-Yourself Projects for Keeping Chickens*

Introduction

Thank you for picking up a copy of *Proven Techniques for Keeping Healthy Chickens!* My name is Carissa and I will be your guide, teaching you the best tips, tricks, hacks, and time-savers when it comes to keeping chickens.

I brought home my first box full of fluff for my birthday after visiting out-of-town friends who had acquired chicks while we were staying with them. I had been toying with the idea of starting my chicken-keeping journey with mature hens, but their fluffy chicks were too adorable to resist! Somehow, I convinced my husband (who is not a fan of spontaneity!) to stop by a chicken breeder on the tail end of the five-hour drive home from their house. A quick jaunt to the farm store and we were set with a box full of peeping babies.

Years later I'm still in love with my chickens. I love that each of my hens has a unique personality. Not only do my girls make me breakfast, but they are fun to watch and even more fun to interact with. I'm lucky that my kids still love the chickens, too, and are my co-conspirators in sneaking adorable balls of fluff home from the feed store (shh, don't tell my husband!).

I am a teacher and mentor at heart. On my website, *Creative Green Living*, and on Facebook in the Creative Green Living Tribe, I teach families how to make healthier choices that are beautiful and delicious—but will also save you time and money and really work! I wanted to take the same problem-solving, time-saving, money-saving, teaching approach I have online and turn it into a book. My friends at Skyhorse Publishing were looking for books about chickens so we had a wonderful opportunity to partner together and bring this book to life!

When describing this project to friends when it was in the development phase, I told them it was like a devotional—but about chickens. If you haven't ever been a part of a mainstream Christian church in America, you may not be familiar with the idea of a devotional. A devotional is a short book, designed to be read just a page or two at a time. You open it up, read a page or two about one specific topic, get a nugget of truth to help you out, and then you go about your day. They are not exhaustive theological treatises or long, detailed commentaries covering every possible approach to a topic. But they are helpful, and people—especially busy people—like them.

This book is designed to work the same way. While you are definitely welcome to read more than one tip, trick, or hack each day, I wanted this book to be easy to digest in small bites. This is not an exhaustive commentary or handbook on every aspect related to chicken keeping. Instead, I'll teach you my best tricks to save you time, effort, and money one topic at a time. The nature of this book is to provide a broad overview or just enough information to get you moving in the right direction. If you are craving a more in-depth treatment of a specific topic, check out pages 213 through 215 for some great resources on plumbing the depths of chicken-keeping knowledge.

I also feel it's worth mentioning that this book really focuses on well-chicken keeping. I am not a veterinarian. While I cover lots of the preventative measures I take to keep my girls healthy, I cannot teach you how to perform bumblefoot surgery or give injections. This book also focuses primarily on keeping laying hens. I touch a bit on roosters, but most of the tips will help you keep your laying hens happy for a long time—as opposed to keeping your meat chickens happy for the eight to twelve weeks before you butcher them.

Thanks again for picking up a copy of this book. I hope you learn something new that makes your chicken-keeping journey more enjoyable and less labor-intensive than it was yesterday!

Glossary of Terms

I throw several common chicken-keeping words around like confetti. If you happened to miss the explanation of what that term means in the description of a previous tip, this might leave you feeling a bit lost. Turn here if you see me talking about something and you have no idea what I mean!

Angry pancake: A term affectionately used to describe a broody hen who snarls at anyone who bothers her (the angry part) and has spread herself out to try and be as flat as possible (like a pancake).

Bloom: The invisible coating around the outside of an eggshell that protects it from bacterial contamination. The bloom generally remains intact until the egg has been washed or cracked. If an egg has been washed (thereby removing the bloom), the egg should be refrigerated.

Brooder: The box or container you use to raise chicks in without the help of a mother hen.

Broody: Depending on the context, broody can be either an adjective or a noun. As an adjective, it describes the state a hen is in when she is either trying to hatch eggs or is raising baby chicks. As a noun, broody is used as short form for "broody hen" or "hen that is brooding."

Chicken math: A condition that strikes most chicken owners when they go from wanting a small flock to suddenly owning many, many more chickens than originally anticipated. If you insist you only want six chickens, but somehow ended up with twelve, you've become a victim of chicken math.

Coop: The small house or shelter your chickens roost in as well as lay eggs in.

Freezer Camp: A slang way to say you are going to butcher your rooster. "I'm getting ready to send my rooster to Freezer Camp" means you are getting ready to kill and butcher him.

Glyphosate: A chemical herbicide designed to kill almost any grass or broad-leafed plant it is applied to. Glyphosate is the active ingredient in Roundup® and other popular commercial weed killers. The use of glyphosate on food crops became especially controversial after the International Agency for Research on Cancer issued a report in March of 2015 identifying it as a "probable carcinogen."[1]

GMO: Short for "Genetically Modified Organism." Refers to any plant or animal that has been genetically modified to contain one or more genes not normally found in that organism.

Organic: For the purpose of this book, organic refers to something certified USDA organic.

Vent: The opening through which a hen's poop and egg both pass to exit her body. Fun fact: Chickens don't pee; their liquid waste is expelled together with their solid waste through the vent.

---

1   Pg 1, "IARC Monographs Volume 112: evaluation of five organophosphate insecticides and herbicides," International Agency for Research on Cancer. March 20, 2015 (http://www.iarc.fr/en/media-centre/iarcnews/pdf/MonographVolume112.pdf)

Choosing Chickens

You've already bought this book, so chickens are on your mind! Whether you love the idea of a self-sufficient pet or want to opt out of the factory-farming system, there are lots of great reasons to choose chickens. For me, chickens seemed like the next logical step beyond my large 400-square-foot garden (a.k.a. "the hundredth acre farm") to gaining control over the quality of food I was feeding my family. Having been obsessed for many years with unusual colors of food (purple carrots, pink tomatoes, and the like), I also loved the idea of being able to collect a rainbow of different eggs every day from hens living in my own backyard.

Check out these next few tips to help you choose the right chickens for your flock. I even put a helpful worksheet in the Eggs-tra Resources section (page 211) to help you keep track of what kinds of chickens are on your wish list.

1
Babies or Big Girls?

One of the first things you need to decide before starting with chickens is if you want to start with chicks (babies) or pullets (big girls). The choice is really up to you and how you prefer to raise your flock.

Benefits of babies

Chickens raised as chicks with human parents are usually friendlier than chickens who come to you as older pullets. They bond with you and even as adults will recognize you and come running for treats or snuggles (or both!). Plus, it's hard to resist how absolutely adorable they are. Children can benefit from learning to raise and care for a baby animal that is dependent on them.

Benefits of big girls

The chief advantage of acquiring chickens that are already mature or approaching point-of-lay is a shorter wait before your girls start laying eggs! Even if a hen you buy has already been laying, she may take a break after moving because the stress of relocating can disrupt her laying cycle. The wait to resume laying after a move will seem like nothing compared to the four to eight months (and sometimes longer!) that you would need to wait to go from a tiny chick to a mature, laying hen. If you are anxious for your chickens to start putting breakfast on the table, starting with big girls might be the right choice for you.

Downsides to consider

If you are raising chicks in your house, know that they will be totally dependent on you for eight to twelve weeks before they are ready to move outside (depending on the weather). If you have plans to go out of town while they are still living in the brooder, you will need to get a chick sitter for them. The other downside to starting with chicks is having to wait for several months before your hens will be mature enough to start laying eggs.

For pullets, it's also good to keep in mind that the older the hen, the fewer eggs she will lay for you. Most hens' peak egg-laying time is in the first two years of their lives. If you buy a two-year-old hen, she should start laying eggs for you sooner than a chick but may not lay as many eggs total over the span of time you have her.

2
Enjoy a Rainbow

Some people crave uniformity while others seek a more colorful life. I love color, and the eggs my flock lays reflect that. To enjoy a rainbow of eggs from your hens, try these breeds:

- **White:** White leghorns, brown leghorns, anconas, and California white chickens all lay white eggs.
- **Cream:** Light brahma, Swedish flower, and salmon favorolles lay eggs that are a creamy not-quite-white but not-quite-brown color.
- **Light brown:** Many chickens lay light brown eggs, but Plymouth rocks (sometimes called barred rock), orpingtons, wyandottes, and Rhode Island reds are some of the most popular breeds that lay light brown eggs.
- **Dark brown:** Black copper marans, welsummer, and penedesenca chickens lay chocolately brown eggs. See page 11 for more about these.
- **Blue:** Ameraucana, araucana, and cream legbar chickens each lay blue-colored eggs. The blue-egg laying gene is similar to the gene a robin carries. See page 9 for more information about blue-egg breeds.
- **Green**: Green egg-laying chickens are called "olive eggers." These are chickens that are a mix of breeds that combine blue egg-laying genes and brown egg-laying genes to create chickens that will lay their own unique shade of green egg.
- **Rainbow surprise:** If you want to play the rainbow egg lottery, Easter eggers (sometimes called Americanas) are fun to have around. These can lay eggs that are pink, green, blue, white, cream, or brown—you won't know until they start laying! Individual chickens will always lay the same colored egg, though.

The eggs featured in this photo, clockwise from the speckled egg, are from: black copper Marans (this was one of her first eggs, which explains the speckled appearance), buff orpington, Easter egger, mille fleur leghorn (it's tiny because it was her first egg), and white leghorn.

3
High-Production Breeds

People have lots of reasons for choosing chickens. For some, it's knowing where your food comes from. For others, it's the novelty of rainbow eggs. If your purpose for keeping chickens is to have lots and lots of eggs, you may want to choose one of these high-production breeds.

- **Australorp:** Australorps became popular in the 1920s after the breed broke several world records for number of eggs laid. An Australorp also holds the world record for most eggs laid in a year for laying 364 eggs in a 365-day period. You can expect a home-raised Australorp to lay five to six eggs a week.
- **Golden Comet:** Also called *gold sex link* or *red star*, the golden comet chicken is a cross between the Rhode Island red and the leghorn, both prolific egg layers in their own rights. You can expect 250 to 300 eggs per year from a golden comet, but one farm I found claims their golden comets lay as many as 330 eggs a year!
- **ISA Brown:** Though not technically a breed, the hybrid ISA browns are bred to be egg-producing machines. ISA browns can lay up to 300 brown eggs per year (about six per week) and are often the egg layer of choice in Australian egg-production operations.
- **Rhode Island Red:** Developed in New England in the late 1800s, the Rhode Island red chicken is also the official state bird of Rhode Island. You can expect a Rhode Island red to lay about 250 brown to pinkish brown eggs a year or about five eggs per week.
- **White Leghorn:** In my personal flock my leghorn, Elsa (the chicken who appears on the cover of this book) lays more eggs hands down than any other chicken I own. Easy and inexpensive to acquire, you can expect a white leghorn to lay about 280 eggs a year or about five eggs a week.

Keep in mind that even with high production types of chickens, lots of factors contribute to the amount of eggs an individual chicken will lay each week. Contributing factors include the amount of light a chicken is exposed to, the nutritional quality of their food, their stress level, and if they are molting. Your mileage may vary depending on the circumstances your flock finds itself exposed to.

4
How to Get the Blues

Non-chicken owners are sometimes surprised to learn that eggs don't just come in brown and white—but a whole rainbow of colors including blue. Similar to the gene that robins carry that gives their eggs the distinctive *robin's-egg blue* color, some chicken breeds will consistently deliver beautiful blue eggs for you.

While there are some myths that claim blue-shelled eggs are more nutritious or have lower cholesterol, neither of these things is true. Regardless of the color of the shell, the nutritional quality of the egg will always be determined by what the chicken that laid it ate. Eggs with blue shells also taste the same as any other colored chicken egg and are indistinguishable once they are cracked open into a bowl.

Blue-egg-laying breeds

If you can't wait to get add beautiful blue eggs to your basket, try these chicken breeds:

- **Araucana:** This unique breed originated in Chile. They can generally only be purchased from specialty chicken breeders, as they have a lower hatch rate than other breeds, which makes them undesirable for commercial hatcheries.
- **Ameraucana:** Not to be confused with "Americana" (a.k.a. Easter egger) chickens, this breed was developed in the United States in the 1970s. Descendants of araucanas, they keep the distinctive blue-egg gene without the hatch rate issues associated with the Chilean breed.
- **Cream Legbar:** Developed in Great Britain in the 1950s, this breed almost died out in the 1970s because blue eggs were not as popular as they are now. They have the distinct feature of being auto-sexing—meaning that male and female chicks look very different from each other so they can be separated by sex with 100 percent accuracy upon hatching. Their eggs will always be blue, but the shade of blue may vary from a very pale blue tint to a greenish teal-blue hue and anything in between.

Easter Egger (sometimes called Americana): Often described as *mutt chickens*, Easter eggers do not belong to any a true breed recognized by official poultry breeding standards. While many Easter eggers do lay blue eggs, they can also lay pinkish, green, cream, and even brown eggs, so if you want a guarantee of getting blue eggs in your basket, they aren't the best choice. They are usually much less expensive than the pure breeds mentioned above, though, so if you are looking for blue eggs on a budget, you might find them worth the gamble.

5
Chocolate Eggs Don't Come From Bunnies

Despite what the Cadbury Crème egg commercials seem to advertise, chocolate eggs do not come from bunnies. They come from chickens, of course! The insides may not be chocolate, but the gorgeous dark chocolate–colored shells range from the deepest brown to a sometimes almost wine-colored egg. If you are looking to add a beautiful contrast of color to your egg basket, choose from the breeds listed below.

- **Barnevelder:** Hardy in cold climates, this breed comes in a huge variety of color patterns and has an unmistakable green shimmer to their dark feathers. Eggs are dark like the black copper Marans, but these super layers will yield about 200 eggs per year. They are friendly and very docile.
- **Black Copper Marans:** The black copper Marans is probably the most common dark chocolate–egg layer. They lay the darkest shades of deep, rich brown eggs. They tend to be friendly, and my black copper Marans hen, Maleficent, made a wonderful mother. Their black feathers shimmer iridescent in the sun and they usually have feathered feet. These beauties lay around 150 eggs per year.
- **Cuckoo Marans:** Cuckoo Marans come in a few different colors with a barred or striped pattern to their feathers. The most common cuckoo Marans are the silver and gold varieties. They lay a dark chocolate–colored egg similar to the black copper Marans; however, eggs tend to vary in color tone and can be speckled. Unlike the black copper Marans, cuckoos have little to no feathering on their legs.
- **Welsummer:** The welsummer is a favorite breed on my friend Melissa's farm. They are similar in appearance to a brown leghorn and lay around 160 eggs per year. Their egg shell color tends to be lighter than the Marans varieties but still dark enough to stand out in your egg basket. Welsummers are great foragers and do well free-ranging.

Regardless of your reason for choosing one breed over the other, you can rest assured these birds will provide you with an abundance of dark chocolate–colored eggs! If you see a breed you want to learn more about, don't forget to add it to your wish list (page 216).

6
Chickens Need Friends, Too!

Whether you decide to build your flock with grown hens or with baby chicks, it's important to remember that chickens need friends. Chickens are communal animals designed to live and thrive in a flock. You should never keep just one chicken by herself. Keeping only two chickens is also a risky proposition because if one gets ill or dies, the other will still be left alone. When buying chicks, I usually overbuy by 10 to 20 percent or at least one extra chick so that if a chick falls ill or dies, you still have the intended number of chickens by the time they are ready to start laying eggs.

Chickens needing friends is also a great excuse to give your friends for why you have so many birds when chicken math sets in. *Chicken math* is a phenomenon observed by chicken keepers where the size of your flock always seems to grow. You might *think* you only want a six-bird flock, but if you have ever heard about a new breed that you just have to have, a friend needs someone to take some extra chicks, or a trip to the feed store for dog food resulted in chicks coming home with you—you may have become the latest victim of chicken math.

Knowing that the chicken math phenomenon is a real struggle for many chicken keepers, do some preventative homework and check out the local chicken ordinances for the town or county you live in. If your local ordinances allow you to keep a dozen chickens, you may want to start with only six so that as your flock grows and chicken math happens, you don't find yourself in trouble with the local authorities. If your starter flock of a dozen birds jumps to twenty-four, your local compliance officer likely won't be persuaded you should have so many friends for your chickens when you hand him a copy of this book to explain why you needed them!

Raising Babies

There is something special about raising baby chicks. The tiny balls of fluff are adorable and heartwarming. You might find yourself surprised at how attached you get to these little critters. Even if you decide to start your first flock with older pullets, I think raising at least one batch of baby chicks indoors is worth the effort just to experience the joy you get from helping a tiny ball of fluff become a full-fledged chicken that lays eggs.

If you have children, the immense amount of learning that happens about chickens in just the few weeks it takes the raise them is impressive. Take lots of pictures and encourage your children to chart their chickens' progress with drawings and observations, to make raising babies an immersive experience.

7
Where Do Baby Chicks Come From?

Where do baby chicks come from? Eggs, of course! But how do they get to you? That's another question entirely! When it comes to acquiring baby chickens, most chicken keepers have a few options.

- **Hatch them yourself:** Using either an incubator or a broody hen, you can hatch baby chicks from eggs at home. Use fertilized eggs from your own flock, order eggs online, or pick up eggs from another local chicken keeper. Learn more about hatching eggs on page 19 and page 20.
- **Buy them at the feed store:** Picking up chicks at your local feed store is an easy way to one-stop shop. In the same day, you can pick up all your brooder necessities and baby chickens. Feed stores often have a schedule of what kinds of chicks they will have available on which days, so be sure to ask about it if you have a type in mind.
- **Order them from a hatchery**: If you don't live near a feed store that carries chicks or you have varieties in mind that aren't available locally, a mail-order hatchery is a good option. Keep in mind that to help the chicks maintain the appropriate body temperature in transit, you may be required to order a minimum number of chicks. Ask a friend to go in with you on your chick order if you need fewer chicks than the ordering minimums will allow.
- **Local chicken keepers:** Many local chicken keepers hatch the fertile eggs laid by their own backyard flock and sell the babies as a small side business. To find people in your area selling chicks, check websites like Craigslist or Next Door. You can also connect with local chicken keepers on Facebook, although be advised that as of 2017, Facebook no longer allows the sale of animals to be advertised in marketplace groups.

No matter where you get your chicks, do your best to have your brooder set up ahead of time so they have somewhere warm and safe to go right away once you get them home. See page 25 for more information on setting up your brooder.

8
Use a Broody to Raise Babies

Has your hen been refusing the leave the nesting box all day? Does she squawk and peck if you try to take her eggs or make her move? If so, you may have a broody chicken on your hands! Instead of being annoyed that this hen has decided to stop laying eggs so she can hatch chicks, embrace this opportunity as a time to raise fresh babies without all the fuss of an indoor brooder.

Help your broody hen become a mom

To help your broody become a mom, you have two options: You can let her hatch the eggs herself or you can trick her into adopting purchased chicks.

To help her hatch her own babies, let her sit on fertilized eggs. Be sure to mark the eggs you are trying to hatch with a pencil or indelible pen. This is because broody moms are known to steal other hens' freshly laid eggs to add to their clutch and you won't want to mix up the two. After about three weeks of sitting with only one or two breaks a day, your fertilized eggs will hopefully hatch!

If you'd like to help your broody become a mom via adoption, you should allow her to brood and sit on dummy eggs for about three weeks. After three weeks of letting her sit on dummy eggs, go into the coop in the middle of the night and swap out the fake eggs with baby chicks. You'll know the potential mama has accepted the babies as her own if you hear her cooing at them and see her tucking them under her feathers. Some mother hens are not so easily fooled so if you see signs of rejection like pecking or bullying, be prepared to collect them to raise indoors in a brooder.

However she got them, once she has her chicks, your mama hen will take the lead, keeping the babies warm and teaching them how to look for food. Be sure to make chick crumble and water easily accessible to the mom and her brood, and in a few months, you'll have new egg-laying flock members that mom will have taken care of integrating herself. Thanks, chicken mama!

9
Hatch Eggs at Home in an Incubator

Whether you are looking for an interesting learning opportunity for your kids or a budget-friendly way to acquire more chickens, incubating eggs is a fun and educational experience. It's also easy as long as you follow a few crucial steps.

The two most important things to pay attention to when incubating eggs are humidity and temperature. Make sure to purchase a combination thermometer (digital is best) that reads both temperature and humidity. This is important because the one that comes on or with your incubator is not always accurate. The incubation period is about 21 days. Eggs should incubate at 99 degrees with 40 percent humidity the first 18 days. For the last 3 days of incubation, humidity is raised to around 60 percent to help aid in the hatching process.

How to incubate eggs

1. Purchase your desired style and size of incubator.
2. Add water to the incubator as shown in the manufacturer's instructions, and allow the temperature and humidity to come up to appropriate levels before adding eggs.
3. Place the fertilized eggs on their side in the incubator.
4. Cover the incubator and do not open it for 3 to 4 days. However, you should monitor to ensure temperature and humidity stays between 99 and 101 degrees with humidity of 40 to 50 percent.
5. After the 3 to 4 days, gently turn your eggs over three times daily until day 18. You can mark one side of the egg with a pencil to help you keep track of which ones you have turned and which side should be up.
6. Try taking a look inside your eggs with a process called candling (read more about this on page 23). Any time after day 7 and before day 18 is a good time to candle the eggs. Be sure not to candle after day 18 as the eggs need to be left alone in lockdown until they hatch.
7. On day 18, increase the humidity to 65 percent. This is also the day the incubator goes into what is called *lockdown*. This means the incubator is not to be opened unless there is an emergency, as the chicks could begin hatching at any time. Opening the cover will release valuable moisture needed to aid in the hatching process.

8. Between day 18 and day 21, you may hear chirps and peeps, and see pips. A pip is a small crack made by the chick's beak as it begins to hatch. Pips can be visible for a full 2 to 3 days before the chick hatches. During this phase, try clucking at your eggs or play them a video of a mother hen—you just might get an answer from the chicks in the eggs!

Now sit back and watch your little fluffies pop like popcorn!

For more detailed information and darling illustrations about hatching chicks at home, I love the book *Let's Hatch Chicks* (Quarto, 2018) written by my friend, Lisa Steele.

10
Look Inside Your Eggs

Sometimes you want to check out what's happening inside an egg. Is there a baby chick growing in there? How is it doing? You can't crack the egg open to look, but how else are you going to see through that shell? Candling to the rescue!

Candling is a good way to see if there is an embryo that is viable in an egg. This is helpful if you found a surprise clutch of eggs and you aren't sure if they have started to develop. It's also helpful if you are incubating eggs so you can remove any dud eggs to prevent them from spoiling. A rotten egg in the incubator can ruin the other healthy eggs.

Before candling eggs, be sure to wash your hands and handle the eggs gently to avoid causing trauma to the embryo. To avoid unnecessary stress on the embryos, try to candle no more than three times between days seven and eighteen of incubation.

What you need to candle eggs:
- A small flashlight or candler
- A dark space like a closet

How to do it:
Using a flashlight small enough that the entire egg covers the end the light comes out of, place the egg on the lit flashlight. If an embryo has started to develop, you should be able to see veins and, depending on the age, you may even see the embryo moving! If the egg is golden yellow and clear inside with no veins, it isn't developing. This is good news if you were hoping to eat the egg and bad news if you wanted to hatch it. Note: You should not eat eggs that you have started trying to hatch even if you eventually find that the egg did not develop. Any "dud" eggs like this should be discarded.

In the case of eggs you are trying to hatch, a great day to candle your eggs is around day ten. At this point, the embryo is large enough to take shape and you can usually see it moving! If you have children, they may like to see the embryos move as well. This is a fun and interesting way to get your children involved in the incubation process.

11
Brooder Essentials

To raise baby chicks inside, you will need a few essential supplies to keep everyone warm and happy. Everything should be easy to find locally if you don't already have it at home, but you can also order online just about anything you need!

- **Box or large container:** This can be a homemade wood box, a plastic tote, a kiddy swimming pool, or a metal stock tank. Some families who have an extra, rarely used bathroom will even set the chicks up in the dry bathtub. Just don't use cardboard because it will absorb moisture, which could start to rot or mold and make your babies ill.

- **Bedding:** I prefer pine shavings and compressed wood pellets designed for livestock. You could also use shredded paper. Don't use flat sheets of newspaper alone, as it's too slick and won't help babies develop proper ankle strength. Also avoid cedar shavings, as the oils in the cedar can be bad for their lungs.

- **Thermometer:** Baby chicks have specific temperature requirements so a thermometer will help make sure everything is correctly adjusted. I like using a wireless indoor/outdoor thermometer. I leave the outdoor temperature sensor in the brooder with the chicks in the warmest part of the brooder and can have the monitor in a different part of the house with me for remote monitoring.

- **Food:** You will need chick crumble (read more about choosing the right food on page 75) as well as a dish for the food. While just about any dish will do, I like the feeders that screw onto the top of a mason jar. The feeder attachment helps keep the chicks from standing (and pooping!) in their food and it's easy to take the whole thing apart and run it through the dishwasher.

- **Water:** To prevent an accidental chick death via drowning, the water dish should be shallow. Just like the feeder, you can get a baby chick waterer attachment that screws onto a mason jar. I like to use apple cider vinegar and garlic in their water. Read more about that on page 89.

- **Heat source:** This can be a heat lamp with an infrared bulb, a chick heating plate (these look like a small table and the chicks go underneath it to get warm), or a chick mat. Read more about how to set up your heat source on page 26.

12
How to Set Up Your Brooder

Setting up your brooder is easy. The goal is to keep your chicks warm, fed, and happy, but these tips will help make things go a little smoother.

- **Heat source:** Whether you are using a lamp, chick plate, or heating mat, it's important that your brooder has different temperature zones. The chicks will decide on their own where they want to be in order to regulate their own temperatures. For this reason, I like to put the heat source to one side of the brooder rather than in the middle so there is a warm side and a cooler side. Put your thermometer in the warmest spot and make sure the warmest area approximately matches the temperature for the age of the chicks based on this chart:

Chick Age	Warmest Spot in the Coop
Less than 2 weeks	90–95°F
2–3 weeks	85–90°F
3–4 weeks	80–85°F
4–5 weeks	75–80°F
5–6 weeks	70–75°F

- **Food:** Set up the food away from the warmest spot. I like to choose an area about halfway between the warmest and coolest areas of the brooder.
- **Water:** A common problem with water placement is that chicks can be messy and will often kick their bedding into the water dish. To help prevent this, either raise the water dish up by putting it on top of a wood block or an upside-down plastic container. Be sure to check on your chicks a few times throughout the day to make sure their water dish is clean and litter free. To learn more about chick water additives, check out page 89.
- **Cover:** It won't be long before your little chicks are not-so-little chicks and want to start exploring the world outside of their box. To help keep the chicks inside the brooder, you may want to get a cover for the box. The cover should allow plenty of circulation so hardware cloth or chicken wire are great cover choices. Don't use covers like blankets or cardboard because they trap too much heat and will cause the chicks to overheat.

You can also add optional things to your brooder like toys, a dye-free, natural feather duster (as opposed to synthetic one), or a dish to serve treats in. If you are feeding them only chick crumble, you will not need to supply grit. If you want to feed them treats or give them little clumps of dirt from your yard to scratch through, be sure to add a dish of chick grit as well to aid in their digestion.

13
Teach Your Kids How to Hold Chicks

Tiny balls of chicky fluff are hard for kids to resist! Regardless of your child's age, if they haven't had experience with chicks, they will need a special session in which you teach them to properly hold a chick in a way that is safe for both them and the chick. Keep chick-holding sessions under 15 minutes.

- **Wash up**: Before handling chicks, have your kids wash their hands. This will reduce the germs and pathogens they introduce to the babies. Make sure your kids wash up after handling chicks, as well. Some chicken diseases can be passed to humans so you want to be sure your hands are clean once you're done spending time with the babies.

- **Sit down:** Sometimes chicks jump or move unexpectedly. Have your child sit down before holding a chick to avoid fall injuries that can happen if a chick were to jump from your child's hands while they are standing.

- **Be gentle**: Unlike a stuffed animal, baby chicks do not like to be squeezed. Squeezing or squishing could seriously injure a chick and result in death. Show children how to use one or two fingers to gently stroke the chick's head or body. Small children may do better with you holding the chick and them petting it.

- **Scoop from the sides:** Teach your kids to approach a chick by putting one hand on either side of it to hold its wings close to its body and then pick it straight up. Never pick chicks up by their heads, feet, or wings.

- **Elevator lift:** Alternatively, you can teach them to lift a chick by putting their open hand on the floor of the brooder. They wait for the chick to walk onto their hand, use their other hand to gently create a cup on top of the chick so it doesn't flap, and then lift their hands straight up out of the brooder.

- **Cupping:** While holding the chick in one hand, use the other hand to create a gentle cup over its back as shown in the picture. This helps keep their wings from flapping and helps the baby feel snuggly and secure.

- **Adult supervision:** Teach your children not to hold chicks unless you are there to help to prevent having chicks accidentally crushed by enthusiastic but well-meaning kids. Encourage them to sit near the brooder and chat with their chickens or to place their clean hands inside the brooder so the chickens get used to them if you can't be there to help them hold.

14
Do a Daily Chick Check

When using a brooder to raise baby chicks, a daily "chick check" is essential to make sure everything is going well. Keep an eye on everything to make sure any problems are addressed before they get out of hand.

Take a peek in the brooder a few times a day (at least twice) to make sure everything is in order. The babies should be happy and moving around unless they are napping. If a chick appears to be dead or lethargic, it may just be sleeping. Try gently touching it to see if it pops up or if it continues to lie still. Lethargic chicks should be removed right away and placed in their own private sick bay (see page 143 for more information on setting up a sick bay).

Check the food and water levels. Baby chicks should have free access to as much food and water as they would like so be sure to refill containers that are running low. Water dishes especially have a tendency to get clogged with litter and other debris so clean out clogged waterers when doing your check.

It's also a good idea to gently pick up each individual chick to check that its eyes are clear and not goopy. You also want to take a look at each chick's bum every day to make sure it is clear and that they haven't developed pasty butt. Pasty butt (also known as *vent gleet*, *pasting up*, or *pasting*) is exactly what it sounds like. If you find any chick that has a "paste" plug of poo stuck to either its vent or the downy feathers near its vent, this is an emergency that is quite manageable but must be taken care of right away. Leaving alone a chick with pasty butt could kill it, as the plug will prevent it from properly eliminating waste. To learn more about pasty butt prevention and treatment see page 32.

15
How to Prevent and Treat Pasty Butt

Pasty butt is an issue that can go from just ugly to deadly, so it's important to keep an eye on your chicks and check them every day.

What is pasty butt

Pasty butt (also known as *vent gleet*, *pasting up*, or *pasting*) is exactly what it sounds like. If you find any chick that has a "paste" plug of poo stuck to the downy feathers near its vent, you have a problem. Be sure not to be confused with an umbilical cord stump in very young chicks—this appears as a tiny black or brown spot about halfway between its vent and belly. If you are unable to see the chick's vent or butthole, that's when you know you have a pasty butt problem.

Treating pasty butt

Although a plug covering a chick's vent is an issue that needs to be addressed right away, you should not just rip off the plug. The dried poop may rip the delicate skin of a chicken's vent or rip out some of the downy feathers near its vent and cause a whole extra set of problems.

Instead, you should soften the plug by holding a warm (but not hot!) washcloth to the chick's bum for a couple minutes. Then use the washcloth to gently wipe away the droppings until they are all removed and her vent is clear. If the plug is particularly stuck, you can also try holding the chick's backside under warm (but not hot!) running water until the plug has been loosened and can be wiped away.

If your chick has been plugged up for very long, clearing the paste away may cause her to suddenly discharge a large quantity of poop, so work over a covered surface to be prepared for that possibility!

Preventing pasty butt

If your chick needed to have a paste plug removed, you can help prevent pasty butt from recurring by gently applying coconut oil or olive oil around her vent with a cotton swab once the plug has been cleared.

Many chicken owners also add apple cider vinegar (the type containing *the mother*) to the chicks' water, which can help prevent pasty butt, as well. If multiple chicks in a batch are suffering from pasty butt, it could be a sign of stress.

If this is the case, give electrolytes in their water for a day before returning to apple cider vinegar water. Read more about different water additives on page 89.

16
How to Help a Lonely Chick

Despite your best efforts to make sure your chickens have friends (see page 13 for more on that!), sometimes a chick finds itself alone. Maybe it got sick and needed to be separated from the rest of its brooder mates. Maybe it is at the bottom of the pecking order and, despite being in the same container with more chicks, it *feels* lonely.

Identifying a lonely chick

A chick who feels lonely will constantly peep for its mother. Even if it's never had a chicken mom that it's known, chicks instinctively know to cheep to call mom to come find it. Whether you are a man, woman, or child, if your chick has bonded with you, to the chick *you* are its mother and it will want you to come save it from its loneliness.

How to help

Oftentimes, lonely chicks can be appeased by being carried or snuggled by a human. Humans, unfortunately, have things like jobs and school and chores so they can't snuggle a lonely chick all the time. Even if you enjoy snuggling your chick to help it not feel lonely, you may need to create a fake mom or flock mate to help appease it while you tend to your human responsibilities.

Some tried-and-true solutions for lonely chicks include:

- Hanging a dye-free, natural feather duster from the top of the brooder so chicks can snuggle inside of it.
- Placing a small stuffed animal inside the brooder to snuggle with.
- Putting a small mirror inside the brooder so when the chick sees its reflection it thinks it isn't alone.

Some chicks prefer the companionship of a feather duster over a mirror or stuffie or vice versa, so if you try something and it doesn't work, keep trying different solutions until you find a way to keep your chick happy. Of course, if your chick is lonely because it is the only chick living in the brooder, you may want to consider purchasing a friend or two for it to live with.

17
Kids and Chickens: Matchmaking Tips

If your kids are like mine, they will love hanging out with their new feathered friends. Help set up your kids and chickens for success with the following tips.

Involve your kids from the start

If you are starting with chicks, the young chickens will imprint on the family members who care for them as babies. If you buy chicks from a mail-order hatchery, try to have your children present when you open the box to help encourage the imprinting process. If you buy your chicks locally, bring your kids to help pick out the birds and set them up in their brooder at home.

Give your kids easy brooder-keeping chores like refilling the feeder or participating in the daily chick check (see page 31). Giving your kids the chance to interact with the chicks daily will help them bond with your kids, so that even as adults the chickens will recognize them.

No boys allowed

While sweet, non-aggressive roosters do exist, thousands of children every year are attacked by roosters in their home flock and some are seriously injured. While hens also have the potential to be aggressive, roosters are much more likely than hens to attack and injure a child. If you have small kids at home, consider a "no boys allowed" flock where you only raise hens and rehome any accidental roosters you end up with.

Give them chicken chores

If your kids are the ones regularly doing things the chickens like, such as letting them out in the morning, feeding them, or giving treats, they will maintain a positive association with your children. My eight-year-old son calls himself "the chicken king" because the chickens flock to him whenever he walks outside. He is involved in their daily care and they know good things will happen when they see him coming!

18
Integrating New Chicks into the Flock

If you already have an established flock of hens, adding new brooder-raised chicks can present a unique challenge to the already-established pecking order. If your chicks were raised by a broody hen, this is usually not an issue, as the mama hen takes the responsibility for socializing the chicks and integrating them into the flock herself.

Warm weather outings

Once your chicks are six weeks old or more, they are ready for some extended outdoor outings on days when the weather is nice. Allow your chicks to be in the same general area as the rest of the flock but protected by a separate cage. I like to use a rabbit run that has sides and a top for this, but you could also make your own protective barrier with hardware cloth or chicken wire.

Place the babies outdoors inside their protected area and supervise to make sure nothing is going wrong. This will allow the older flock members to check them out and be aware of their presence without feeling intimidated.

Sleepovers

If your coop is large enough and the overnight temperatures inside your coop are warm enough, you may want to add a separate chick sleepover area to your coop. This could be a large wood box or feeding trough inside the main coop with hardware cloth or chicken wire over the top to keep little chicks in and big chickens out. Move the chicks out to this box with their food and water. This will help the older birds get used to having the smaller ones in their space without having to interact directly. Don't start doing sleepovers until the chicks are weaned off their heat lamp, as heat lamps should not go in the coop.

Overnight switch out

Some people have had good luck with flock integration after doing some warm weather outings by placing the juvenile chickens on the roosts inside the coop overnight while the older chickens are asleep. Waking up together seems to cause fewer issues than if an older bird saw a young one strut right into the coop, which could be perceived as a threat or invasion. If you try this, try to be awake before dawn to peek in on them and make sure this is going smoothly.

In the Hen House – All about Coops, Runs, and Shelters

♥ ♥ ♥

Eventually your chickens will need to move out into their own space. Whether it's a modified playhouse, a prefabricated coop in a box, or a custom-built chicken palace, most families rejoice on the day the chickens move outside. Keep your hens healthy and happy by following these tips for setting up their outdoor accommodations.

19
Choose the Right Size Coop

To determine the right coop size for you, you first need to determine how your coop will be used. Many chickens live completely self-contained in their coop/run combo. If this is your set up, you will need to have more space in your coop than a flock who only goes in their coop to sleep and lay eggs and spends the day roaming around a large yard or field.

The general rule of thumb is that you need a *minimum* of:

2.5 square feet of space per bird if they free-range during the day

and

14 square feet of space per bird if they are always confined

Keep in mind that these are really minimum numbers and chickens are always happy to have more space than less, so be generous and round up when doing space calculations. Also keep in mind when choosing a coop that chicken math is a real phenomenon and you are unlikely to always have the same number or fewer chickens than what you first start with. In fact, I would go so far as to say that you should figure out how many chickens you want to have and then build a coop that is *at least twice the minimum size* needed for that many birds. If you don't end up expanding your flock, your chickens will be happy with the extra space, and if you do expand your flock, you will be able to accommodate them without having to build or buy a new coop.

20
Playhouses and Other Chicken-Coop Hacks

One thing I noticed when I was shopping around for chicken coops for my first flock was that prefabricated chicken coops were quite pricey. I could easily spend $600 for a chicken coop large enough for twelve free-range hens that reviews said would fall apart or need significant repair within three years. On the other hand, I could spend $300 for a brand new similarly sized child's playhouse. If I bought a second-hand playhouse, I could spend even less!

I ended up buying two playhouses (my children claimed the first playhouse as their own, causing me to need to buy a second playhouse for the chickens!) and about $50 in additional materials to predator-proof it. Altogether, I spent less than $400 and my kids got a playhouse out of the deal, as well!

To turn the playhouse into a coop, I covered all the windows and other openings with half-inch hardware cloth, added a hardware cloth skirt around the base (see page 133), built a simple foundation with bricks so the wood structure wouldn't be sitting on wet ground, added nesting boxes on the outside of one of the large windows, and added raccoon-proof locks to both the front door and nesting box lid.

If you are having a hard time finding an affordable coop and don't have the building skills to build one from scratch, look around you! In addition to playhouses, you could also do something similar with a shed. I've also seen chicken coops made from old trampoline frames, doghouses, and even an abandoned pickup truck camper!

At a minimum, a chicken coop needs to:

- Be weatherproof
- Be predator-proof
- Provide roosting space
- Have nesting boxes for laying eggs

That's it! That leaves you lots of room to brainstorm and be creative. An afternoon browsing through Pinterest or online chicken-keeping forums is sure to leave you with dozens of ideas for turning your household discard pile into viable housing for your chickens.

21
Hardware Cloth Is Your Hero

Despite its name, chicken wire really has very limited uses in chicken-keeping contexts. The iconic net shape of a piece of chicken wire is flexible and easy to manipulate—which is precisely why it's terrible for predator-proofing anything. One thing that actually does all the stuff people think chicken wire should do is hardware cloth.

Hardware cloth

Unlike chicken wire, hardware cloth is made from welded squares in a grid design. The different sizes of hardware cloth refer to the different-sized squares it can form. I find half-inch hardware cloth to be the most useful for most chicken-keeping applications.

Hardware cloth is durable, which makes it a great choice for predator-proofing chicken coops and chicken runs. Use it to cover windows and vents or to make a skirt around the bottom of your stationary coop (see page 133 for more details on how to do that).

Chicken wire uses

While chicken wire is terrible at predator-proofing things, it does make a good chicken-proofing cover. Use chicken wire to protect plants you don't want your chickens to eat. Or to make a temporary fence to keep them out of a particular area. Just don't expect it to stand up to things any larger or stronger than a chicken!

22
How to Set Up Nesting Boxes

In addition to making sure you have enough coop space for all of your girls, it's important to make sure you have enough nesting boxes for them. If you start with baby chicks, you won't need to worry about having nesting boxes until they are at least four months old. If you start with pullets or hens who are already mature, be sure to have nesting boxes ready to go right away.

How many boxes?
A good general guideline is to have one nesting box for every five or so chickens. Many chicken owners like to joke that no matter how many boxes they have given their hens, they all seem to choose the same favorite nesting box to lay their eggs in.

Chickens often have their own ideas of what an ideal egg-laying spot looks like. I once had a hen who preferred to lay eggs in a plastic garbage can that was holding compost instead of either of the lovely boxes in the coop. Even so, having more nesting boxes than the chickens think they need is good, especially when someone decides to start brooding.

Setting up the boxes
No matter where in the coop they are installed, the nesting boxes should be lower than the roosts or else your chickens might find them to be a nice place to sleep. Since chickens do a significant amount of pooping while they sleep, this is definitely NOT something you want them to do! Yuck!

Boxes should also be filled with soft nesting material. I put a soft liner in mine and then pile a few handfuls of pine shavings on top with a sprinkling of dried herbs (see page 59 for that recipe!).

While I personally find it to be a convenient feature, your nesting boxes don't need to be accessible from the outside of the coop. If they are, make sure to put a predator-proof latch on them so that it doesn't become an entry point for a hungry raccoon or bear.

23
Heat Lamp Alternatives

While heat lamps can be useful (with precautions!) for brooding chicks indoors, the truth is that they are a fire hazard and shouldn't be used once chicks are fully feathered and able to live outside. Even during the winter months when the temperature drops, heat lamps truly aren't needed. Chickens have been surviving in the wild in much harsher conditions than your coop for thousands of years—all without heat lamps!

Heat-lamp-free ways to brace for the cold

Worried that your adult chickens will freeze to death once the temperatures drop below freezing? Prep your coop with these tips to keep them cozy and safe. Checking in with other chicken owners in your area (connect with them on Facebook or in online forums) is another to way to double check that your winterization methods will be adequate based on your local climate.

- **Protect against drafts:** Cover drafts, especially in areas where the cold air could be blowing directly on the chickens. Your coop should still be ventilated—just up high where it won't blow directly on the them. For my coop, I hang thick plastic on the outside of the windows during the winter. This still allows air to get out of the coop but keeps cold air from blowing in.
- **Give them wide roosts:** Make sure your roosts are wide enough for them to comfortably grip. Some people like to use 2 × 4s turned wide-side up or wood designed to be used as hand rails. Thick branches also make good roosts. Once comfortable, they will fluff up their feathers around their feet to stay cozy.
- **Snacks at bedtime:** Digesting food at night keeps a chicken's metabolism going. Some keepers like to feed late-night snacks like corn or other scratch grains before they turn in to make sure the girls have enough calories to burn all night.

Still worried?

Remember, your chickens wear down coats all day long and snuggle up on the roost at bedtime. They will be much happier sleeping outside in the cold than a human would be! Heat lamps are fire hazards that cause an average of 830 fires annually resulting in $46 million in property damage each year! If you take the steps above, you can keep your flock warm without putting them and you at risk.

24
Automate Your Door

When people ask me how I make time to keep chickens, I tell them that the chickens basically take care of themselves! And it's true. One way you can help your chickens be more self-sufficient is with an automated door.

Is automation for you?

Are you up early and around in the evening when it gets dark? If so, letting your chickens in and out on your own might work well for your schedule. If you like to sleep in, tend to get home well after dark, or your weather is often subpar, you might want to consider automating your chicken door.

A properly set up automated door will let the chickens out sooner—which will give them more time to range and more exposure to sunlight. It also makes sure they are locked up before nighttime predators start creeping around. Humans in rainy or snowy areas like them because automatic doors mean you don't have to gear up for a quick trip outside through a storm to lock up your birds.

Kinds of automatic doors

The most popular type of automated door is a small, chicken-sized door that lifts straight up, called a *pop door*. If you converted a playhouse that uses a swing door into a chicken coop, you can still automate your door with either a remote-control motor or an automated one from a company called Add-a-Motor.

Things to keep in mind

If you decide you want to automate your chicken door, keep these tips in mind:

- **Don't forget to check your chickens**: If you decide to automate your chicken door, you should still plan to pop outside to check on the chickens at least once a day. This lets you solve any problems—like if they knocked their water over—and collect eggs left in the nesting boxes.
- **Avoid lockouts:** To avoid one or more chickens finding themselves accidentally locked out of the coop at night, either choose a door automator that gives a warning or set it to lock down thirty to sixty minutes after everyone should be on the roost.
- **Have a backup door:** If one of your doors is automated, it's a good idea to have a second way into the coop in case the door malfunctions.

● **It still needs to be predator-proof:** Not all doors will come with an inherently predator-proof design. Raccoons are able to lift some pop doors straight up, so be sure your door design includes an automated latch or lock that is raccoon-proof. There are lots of great videos demonstrating predator-proof chicken doors on YouTube.

25
Deep Litter

While it is true that there are many different ways you can set up the bedding for your chickens, my favorite method is called *deep litter*. This method is easy and low maintenance—requiring less time to clean every week compared to other litter types like sand. The combination of the carbon from the shavings and the nitrogen from the droppings starts the composting process while the litter is in the coop. Not only does this cut down on composting time once the coop is cleaned out, but the composting process also creates heat. This additional mild heat is especially beneficial during the winter.

How to do it

Start by putting down a layer of pine shavings about six inches deep in the bottom of the coop. As the litter starts to discolor from dampness and mixing with chicken droppings, or if it ever starts to smell bad, add a fresh layer of about an inch of pine shavings to the top. Every so often, add a fresh one-inch layer to the top. I find adding pine shavings once a week works well for my coop and flock size, but you will soon figure out what kind of routine works well for you.

Twice a year (I usually do this in the spring and the very beginning of the fall), clean out all but an inch or two of the shavings and add another four inches of fresh shavings. Move the soiled shavings either to a compost pile or barrel to sit for another six to twelve months. This will allow it to complete the composting process before adding it to your garden beds.

Tips for deep litter

Be sure to follow these pointers to get the best results from your deep-litter coop:

- **Use pine**: Cedar shavings can cause respiratory issues and should be avoided. Don't use hay or straw which can harbor mold and fungus and cause other issues for your flock.
- **Add a threshold**: Adding a small piece of wood across the door of your coop will help keep the litter from spilling out when you open the door.
- **Watch for odor:** Strong odor, especially an odor of ammonia, means you have too much poop and not enough shavings. Add more shavings to combat this.

- **Check for unstirred pockets**: Generally, your chickens should do a good job keeping the litter stirred, but if your coop is very large, some areas may be neglected and will just need to be flipped with a shovel to keep the mix aerated.
- **Don't use DE:** While diatomaceous earth (often referred to by its initials) is a great tool to have in your chicken-keeping arsenal, don't use it in your deep litter. Beneficial microbes, bacteria, and bugs are part of how a deep-litter system works and DE will inhibit a proper composting process.

26
Poop Catchers

While I think that deep litter is a superior method of managing coop poop, if you can't do deep litter for whatever reason, the second-best way to manage poop is with a droppings board or hammock. The theory goes that since chickens do most of their pooping at night while they are roosting, if you can catch most of that poop before it ends up in the litter, you can keep your coop cleaner and litter changes less frequent.

It's important to note that this waste-management method will require daily work and maintenance. If you aren't ready to make a commitment to daily poop cleanup, flip to page 53 and learn more about deep litter to see if it might be a better fit for you.

If you want to try a poop-catching method, here are three different ways you can do it:

- **Chicken litter box:** Based on a similar idea to a cat litter box, roosts built this way feature a coop bar over a board or bin about eighteen inches wide and filled with sand or a zeolite mix. Every morning, you use a litter scoop to scoop the droppings out of the sand or zeolite and add it to your chicken poop compost pile.
- **Slide-out board:** Some coops feature a side-access drawer of sorts where you can slide out the poop board for easy cleaning from outside the coop. For these types of setups, line the poop board with newspaper or pine shavings so you can dump the whole thing right into the compost pile (sand isn't ideal for this setup because you will end up with too much sand in your compost).
- **Poop hammock:** Exactly how it sounds, this method uses a piece of plastic or other waterproof material such as a tarp to create a hammock below your roosting area. The poop accumulates in this hammock, which is then brought outside and dumped into the compost pile once a day.

27
Make a Nesting Box Spa Blend

Give your ladies a bit of a spa day by creating a nesting box dried herb blend that smells great! Many swear by nesting box herbs to keep pests at bay and the coop smelling fresh. While you can use fresh herbs and flowers with your flock, I like to make a large batch of this dried herb blend and store it for use throughout the year.

Herb blend recipe
To make your herb blend, mix equal parts of any of these dry herbs and flowers and store in an airtight container:

- Basil leaves
- Calendula petals
- Chamomile flowers
- Dill leaves
- Echinacea flowers
- Lavender buds
- Lemongrass
- Parsley leaves
- Peppermint leaves
- Rose petals
- Rosemary needles
- Sage leaves
- Spearmint leaves
- Thyme leaves

The more herbs you use, the more colorful your mix will be!

How to use it
Sprinkle the herb blend in their laying nests or over their bedding for a colorful, sweet-smelling treat. Don't be surprised if you see your chickens sneaking a taste of their custom herbal mix.

If you have fresh flowers or herbs growing in your garden, chickens love to get the fresh versions of these plants, as well. Just about any herb or flower that is edible for humans will be okay to give to your chickens, so don't be afraid to leave the clippings for them as you trim back your plants while gardening.

28
Keep Your Chickens from Flying the Coop

Depending on what kind of coop and run setup you have, keeping your chickens in their designated area might be a little challenging.

Solution 1: Hardware cloth

If your run area is small, covering the top of the run with hardware cloth is likely the simplest solution. This will let in plenty of fresh air and sunlight while keeping predators out and chickens in.

Solution 2: Bird netting

If you have a run area larger than a pre-fab chicken run but smaller than a backyard, bird netting might be a good option. It is effective while still being less expensive than buying that much hardware cloth. Using the established fence as the support, drape bird netting over the top and pull it snug so it isn't droopy. Note that while this may provide some protection from predators like hawks, this is not really a predator-proofing solution against ground-based predators like raccoons or dogs, so it may not be appropriate for all situations.

Solution 3: Wing-clipping

If you have a backyard or field where you range your birds, creating a lid of sorts on their enclosure is not practical. Wing-clipping is an easy and painless way to keep them from hopping your fence and destroying your neighbors' gardens. To clip their wings, hold your bird and extend its wing out to see its feathers. Use sharp scissors to cut the longer flight feathers on just one wing. If you cut both wings, it won't work! With a clipped wing, your birds will still be able to fly a bit, but having the flight feathers on one wing shorter than the other keeps them from flying up very high. This method should keep all but the most stubborn of chickens on the right side of the fence!

29
Keep Your Free-Rangers Safe

If you like to let your flock free-range in your yard or field, it's important to create safety shelters for them. This gives them a place they can run for cover if they see a predator coming, particularly from above. A good cover could be natural or man-made. Here are some safety shelter ideas that are popular with birds and chicken-keepers alike:

- **Trees:** Planting a tree with a drooping top such as a weeping cherry will provide a nice hideout for your chickens. It also doubles as a tasty snack in the spring when they can eat all the flowers that appear at chicken height or lower! Fruit trees like apples, pears, peaches, or plums trained shorter and with a wide canopy are good options, too.

- **Bushes and shrubs:** Bushes and shrubs with an overhead canopy but little nooks at the base where the chickens can squeeze in are great, too. Plants like red-tipped phontinia, butterfly bushes, and blueberries are good for this purpose. Even without predators around, you might find them lounging in the shade of their favorite plants on a regular basis! Of course, if you plant an edible like blueberries, plan to sacrifice anything growing below about thirty inches to the chickens. You can have the berries at the top.

- **Man-made shelters:** If you have a large open space where your chickens will roam but natural shelters like bushes or trees are impractical, you can set up little hiding huts for them. Chickens will instinctively know to run there to hide when faced with a predator and will likely pop in to get a break from the sun or the wind, as well. You could repurpose child playhouses or make a simple lean-to with a couple posts and a weatherproofed sheet of plywood. If you're feeling particularly creative, you could even turn the roof into an herb garden with chicken-friendly plants.

30
Landscape Your Chicken Yard

Just like humans, chickens love landscaping! Plants can have lots of different functions in and around chicken yards. Consider using a variety of plants to improve your chicken space.

- **Vines:** Ideally, the area where your chickens spend most of their time will have a good mix of sun and shade. If the area you are planning to put them is mostly in direct sun all day, try planting an edible vine to climb up the fence on the outside of their run. Climbing roses are popular for this, as the chickens love to eat the flower petals when they fall. Grapes and hops are popular vining plants, as well. If your chickens range in a yard where these plants are growing, be sure to protect the roots from being scratched up by making a barrier with chicken wire or large rocks.

- **Trees**: If you don't have the right setup for planting a vine, trees can also be used as both shade and protections from overhead predators. I have a weeping cherry tree in my yard for this purpose. Non-weeping fruit trees are also good choices and your chickens will be happy to help you clean up any fruit that falls to the ground within their reach.

- **Bushes:** When it's raining, my chickens prefer bushes to any of their other shelter options. Bushes are also great in open-top runs and ranging areas to give protection from overhead predators.

- **Herbs:** Herbs have so many uses in chicken-keeping that it's nice to be able to grow your own. If you have an enclosed run, try growing mint and lavender around the perimeter to repel pests. If your run isn't enclosed, plant the herbs under a chicken wire cloche. This lets the plant have a protected base that won't be eaten and lets the chickens snack on any parts that grow outside the wire.

- **High-fragrance flowers:** Honestly, chicken poop stinks. While there are lots of strategies for helping reduce smell with different types of litter and cleaning systems, one way to make the smell of your chicken yard less offensive is by growing highly fragrant flowers between your chicken yard and your house. Roses, lilacs, gardenia, lavender, and lilies are all wonderful, fragrant options that will be safe for your chickens if they decide to take a bite.

31
Come Home to Roost (and What to Do if Your Chickens Won't)

When changing coop locations, chickens sometimes get a bit . . . lost. Whether you moved your fully feathered chicks outside to live into the big-girl coop or you brought a new chicken into your flock, if the chickens aren't 100 percent sure about where to sleep, they might settle in some place where you don't want them. They will generally start trying to roost in the highest available spot they can get to—even if that spot isn't safe or where you want them sleeping. This might mean you find them sleeping in trees, on cars, on top of the roof of their coop, or huddled together on your porch if you left the light on. There's nothing quite like coming home to a porch full of chicken poop because your girls decided that was the safest place to spend the night!

If your chickens are having a hard time putting themselves to bed in the right place, try these tricks:

- **Put them where you want them:** Pick up your stray chickens who are sleeping elsewhere and put them on the roosts where you want them to sleep.
- **Keep them cooped up:** If you have a large coop that provides for at least fourteen square feet per bird, you could try leaving them locked up for a full day or two (with adequate food and water, of course). This will help them imprint on the coop and identify it as home base. If your coop is smaller than that, you might try leaving them locked up for an extra hour or two in the morning but definitely don't leave them in there all day.
- **Leave a light on:** Chickens are attracted to light, so as the sun is going down, if there is an area that remains lit, they will instinctively head in that direction. Try leaving a battery-powered lantern, a collection of solar-powered garden lights, or an LED nightlight on for them inside the coop to help draw them inside. Note: Do not use heat-emitting lights for this purpose; they are a fire hazard.
- **Block off-limits spots:** If your chickens have managed to make it inside the coop but are sleeping in an undesirable location such as in the nesting boxes, block off those areas. Blockades can be temporary (such as putting empty plastic bins inside nesting boxes in the evenings) or permanent, depending on whether you want them to have access to that area during the day.

32
To Free-Range or Not?

When considering your chicken setup, you need to decide if you will allow your flock ranging privileges all, some, or none of the time. My flock gets free access to my backyard from late September until April or May (whenever I plant the garden). The other five months of the year, they stay in their very roomy open-top run, with occasional supervised outings the last hour before dusk. Some flocks get freedom year-round and some always stay in an enclosed space or only leave it under intense supervision.

To help you decide if ranging in your yard or field is right for you and your flock, consider some of these pros and cons:

Free-ranging pros
Chickens:

- Are less likely to get bored;
- Are less likely to fight with each other;
- Have access to extra nutrients through bugs, grubs, and plants they find to eat;
- Lay more nutrient-rich eggs (see above);
- Require less feed than confined chickens;
- Will deposit free fertilizer in whatever areas they are allowed to roam;
- Will help control pests such as spiders, snakes, and small rodents; and
- Are happier.

Free-ranging cons
Chickens:

- Are more vulnerable to predators;
- Might dig in undesirable locations;
- Could damage plants, gardens, or other landscaping;
- Might poop in undesirable locations;
- May lay their eggs in inconvenient locations; and
- Are more subject to the weather.

33
How to Choose a Coop Location

You may have your birds, you may have a coop or materials to build a coop, but now you need to decide where to put it! As you are looking around and deciding where your new feathered friends should live, be sure to consider these things:

- **Local ordinances:** Be sure to check local laws and zoning ordinances to see if there are rules regarding how far from your property line or neighbors' homes your coop needs to be.
- **Sunshine:** Sunshine is a crucial component for proper egg production. At the same time, you don't want your chickens baking under intense sun all day. Your best bet will be an area that gets good sunlight but also has some shade cast from plants or other buildings during different parts of the day. This will let your birds get out of the heat as needed.
- **Water:** If your birds' feet are constantly wet, that is a recipe for disaster. Chickens need to have a dry spot for both dust-bathing as well as general standing around. Don't put your coop in a low spot that is prone to flooding.
- **Concrete:** If you live in an area where your flock might be subject to burrowing predators who could come from underneath or if your yard is prone to flooding, consider putting your coop on top of a concrete slab for protection and allowing them to range in the grass during the day.
- **Weather:** Consider your local weather patterns when deciding on a coop location. If you live in an area with a strong south wind, for example, you may want to build your coop and run against a fence to serve as a windbreak.
- **Heavy-metal hazards:** Make sure you locate your coop away from any heavy-metal hazards such as lead-painted buildings. This topic is important enough to warrant its own page, so check out page 71 for more information about that.

34
Avoid Lead-Poisoned Chickens

Chickens are very good at searching out sweet things. This trait becomes a big problem if you are keeping your chickens in an area where lead paint is (or was) present. Because lead is sweet, they love how it tastes. Chickens are particularly good at finding tasty lead-paint chips that may have fallen into the dirt years ago. If they have access to a building painted with lead paint (even if it has been painted over with many layers of fresh paint), they will continue to pick at the paint to get to the sweet stuff underneath once they figure out it is there.

Why is this a problem? If chickens eat lead or scratch in lead-contaminated soil, the lead will enter their blood stream and concentrate in their ovaries and any place their body typically stores calcium (like the egg gland and their bones). This, of course, becomes bad news for you because if you eat eggs from lead contaminated chickens, you will be eating lead, as well!

How can you avoid all of this? If you plan to keep chickens in an area that has buildings or fences built before 1980, be sure to screen for lead before moving them in.

How to screen for lead
Be sure to check for lead paint on the outside of any buildings or fences on your property that were built before 1980. You should also check the paint on any farm equipment (regardless of age) that chickens have access to. To help you determine if the paint is leaded, you can purchase a lead-paint testing kit at any hardware store or online and follow the instructions that come with the kit.

In addition to looking for lead paint, you should also have the soil tested for lead in the areas you plan to give your chickens access to. Many counties or universities offer free or low-cost lead testing for soil. If your soil tests positive for lead above 40 PPM (parts per million), move your chickens to a different part of your property that is not positive for lead.

If your chickens are already laying eggs, some labs and universities offer lead testing on fresh eggs for a modest fee. This will help you accurately determine if you are being exposed to lead through your chicken eggs.

What to do if you find lead
If you discover lead paint or lead-contaminated soil on your property, this is not an issue to take lightly. Be sure to locate your chickens away from any lead hazards and block access to buildings or equipment painted with lead paint.

Feed Me! Food and Snacks for Your Flock

🐔 🐔 🐔

They say "you are what you eat," and this saying is just as true for chickens. It's not just that chickens are what they eat—but their eggs are what they eat, as well. If your birds are well fed with nutrient-dense food, they will pump out nutrient-dense eggs for you to enjoy. Use these great tips and tricks for keeping your chickens well fed and happy so that they can keep you well fed and happy too.

35
Choose the Right Feed for Your Chicks

In the beginning, chicks have some pretty specific nutritional requirements, so this is one area where I recommend you stick with a prepackaged chick crumble and not try to DIY their food. That said, you do still have some options when it comes to feeding your babies.

Medicated chick crumble

Medicated chick starters contain a medication used to prevent coccidiosis, which is an intestinal parasite prevalent just about everywhere. The feed will not treat coccidiosis, just prevent it. Honestly, most home chicken-keepers won't need medicated feed to keep coccidiosis under control as long as they are keeping their brooders clean. I also don't personally use medicated feed in part because I have a strong preference for organic options and there are no medicated organic feeds on the market.

Unmedicated chick crumble

Unmedicated chick crumble usually contains about 18 percent protein and is the perfect first food to feed your chicks until they are about eight weeks old and you switch them to grower feed. There are many brands of chick crumble on the market including some that are non-GMO certified and/or organic. Due to the prevalence of genetically modified organisms (GMOs) in the American agriculture system, unless you specifically choose non-GMO or organic feed, your chick feed is likely to contain GMOs. For more information on GMOs and why I don't prefer them for my chicken feed, check out page 77.

Chick grit

While your chicks are only eating chick starter, they won't need grit, as the starter is easily digested without it. Once you start giving treats like worms, grass, seeds, etc. (anything that isn't chick starter), they will need access to grit to properly digest them. Bags of grit are inexpensive at the feed store. To give it to your chicks, just leave a little self-service container for them in the brooder and they will help themselves as needed.

36
Is Organic Worth the Extra Price?

It's hard not to notice the price difference between organic and "conventional" chicken feed (the nonorganic stuff) when you are standing in the store looking at them side by side. In some cases, organic feed can cost up to twice as much as the nonorganic feed! Is the price worth it? That all depends on why you are keeping chickens.

The difference between organic and conventional crops
Currently about 92 percent of the corn and 94 percent of the soybeans grown in the United States are GMOs. Because of this, just about any chicken feed you buy is going to contain GMOs unless it is specifically non-GMO-project verified or USDA certified organic. Most GMOs are designed to either withstand large applications of weed-controlling chemicals like glyphosate (the active ingredient in Roundup®) or contain a bacteria commonly known as BT. Harvested GMOs will contain trace levels of these chemicals and/or bacteria.

Certified organic crops, however, are subject to strict regulations and inspections. GMOs are not permitted in organic agriculture and neither is the use of glyphosate. The USDA Organic program also limits the use of food additives including preservatives or ingredients that might be added to make processing easier.

Should you choose organic feed?
In my home, the vast majority of the food we buy at the grocery store to feed our family is organic. One of the reasons I like keeping chickens is to have control over the quality of our food and to reduce our overall pesticide exposure, so choosing organic feed made the most sense to us.

If your family chooses mostly conventional or prepackaged foods, you are already eating a large quantity of GMOs. If organic food isn't a priority to you and GMOs aren't something you are actively trying to avoid, you might prefer the cost savings you get from purchasing conventional feed.

37
Stretch Your Food Budget with Fermentation

Laying hens eat about a quarter pound of feed per bird, per day. At this rate, even a small flock of only four chickens can burn through a bag of feed quickly. This can start to feel like a financial burden, especially if your chickens are eating a premium food like organic whole grain feed.

Regardless of the type of feed you are giving your chickens, you can help your food budget stretch further by fermenting. Not only is this good for your pocketbook, but the fermentation process increases the availability of the nutrients in the food to your chickens, resulting in happy, well-nourished birds.

How to ferment feed
Fermenting is easy. To start, you will need three containers that can hold a little more than one day's worth of feed for your chickens. For a small flock, this might be a large glass jar. For larger flocks, this might be a bucket.

- **Day 1:** Fill one container a little less than halfway with chicken feed. Fill to within about an inch or two of the top with filtered water and cover loosely.
- **Day 2 and Day 3:** Fill a new container a little less than halfway with chicken feed. Fill to within about an inch or two of the top with filtered water and cover loosely. Place this container in a line behind day one's container (on day three, place the container at the back of the line behind day two).
- **Day 4 and every day after:** Feed the food from the oldest container (at the front of the line) to your chickens. Rinse the bucket and repeat the process from the previous step.

Avoid exploding feed
Did you notice that in the instructions I tell you to leave some room between the water level and the top of the container? This is because, as the feed absorbs water, it may expand and need room to move around. If you fill the container all the way to the top, it may overflow as it expands and leave you with a big mess!

Also avoid covering fermenting containers tightly. Cover loosely to keep out flies, but if your cover is too tight, it might explode or overflow, as the fermentation process creates gas that is trapped by a tight lid.

38
Chickens are Composting Alchemists

Before I was a chicken owner, I was an active backyard composter. None of our food scraps, weeds, or yard trimmings went to waste because I would turn them back into nutrient-rich garden soil via composting. I am still an active composter but how we compost has changed.

Food scraps

Instead of dumping them in the tumbling composter, we feed a vast majority of our kitchen trimmings and leftover food to the chickens (see page 84 for more on which foods are safe to feed to chickens). This not only reduces their feed bill and makes the chickens happy, but it allows the food to be "pre-composted" (a term coined by my friend, Lisa Steele, author of *Gardening with Chickens*[2]) by the chickens. No need to wait weeks for the worms and bacteria to break down that food; the chickens will do it for you in less than twenty-four hours!

Yard waste

Dead leaves, weeds, herb trimmings, edible flowers and finely chopped grass clippings are also great candidates for "pre-composting" by your chickens. Not only are they a tasty snack, but they also provide hours of entertainment for the chickens as they dig through and shred yard waste for you. Move anything they didn't eat to your actual compost pile.

Stuff you still need to compost

While the chickens will "pre-compost" your food scraps for you, you still need to actually compost your chicken droppings and bedding from the deep-litter method (see page 53). Anything moldy or rotting should also skip your chickens and just go straight into your regular compost pile. If you leave the pile in an area accessible to the chickens, you will probably see them sorting through it to pick out any tasty morsels left behind or worms and insect larva that have made the compost home. The turning and fluffing action just increases the speed at which the compost works. Thanks, chickens!

---

2   Lisa Steele, *Gardening with Chickens: Plans and Plants for You and Your Hens* (Voyageur Press, 2016).

39
Make Feeding Time Interesting

Being a chicken can be boring, especially if you are cooped up all day. Getting a little variety at mealtimes helps keep your day interesting. Since interested chickens are not only happier, but less likely to pick on each other, it's a win all around! Try these tricks for keeping mealtime interesting for your flock.

- **Scatter their feed:** If you use a whole grain feed that can stand up to the elements (as opposed to crumble or pellets, which disintegrate when they get wet), you can try scattering their feed around their run or range area. Hunting for food is more interesting than eating it from the feeder.

- **Add treats:** While feeding them a nutritionally balanced feed for most of their diet is essential, chickens always love treats! Popular treats include freeze-dried mealworms, soldier fly larvae, scratch grains, table scraps, herb trimmings from your garden, live crickets, and fresh fruit.

- **Make a snacktivity:** Turn snack time into an activity (what I like to call a *snacktivity*). Try putting their treats in a treat ball or turn a piece of fruit into a piñata (see page 93 for info on how to do that). You can even make mini snack blocks (see below) and hang them around the run to keep things interesting.

- **Add a snack block:** While the official "Flock Block" is a proprietary product you can buy at a feed store, you can also make your own flock-block-style snack if you'd like. Follow the recipe on page 91 to keep your chickens happily pecking at something other than each other.

40
Can My Chickens Eat That?

Chickens can eat just about anything you do and lots of things you wouldn't (garden snakes, anyone?). This makes them great partners in helping you clean out your fridge and pantry and just reducing overall food waste in your home. It also makes them excellent exterminators—I once found a giant spider in my bathtub so my solution was to go outside and catch a chicken, bring it inside, and put it in the bathtub with the spider. Of course, she gobbled it right up and I couldn't have been more pleased!

Similar to humans, chickens each have a particular set of likes and dislikes. My chickens, for example, do not care for carrots (unless they are tiny carrot sprouts growing in my garden) but they will chow down on any other food scraps (and bathtub spiders!) I want to feed them. If you offer something to your chickens but they haven't eaten it after a couple days, be sure to move it to the compost pile so it doesn't begin to mold.

There are a few things you should definitely NOT feed your chickens, though. If you are ever in doubt about whether a food is chicken-safe, a quick Google search can usually help you confirm before you set something out for them.

The never list
You should never feed your chickens any of the following:

- Avocado pits or skins
- Green potatoes
- Raw potatoes or potato skins (cooked potatoes are okay as long as they aren't green)
- Unripe tomatoes
- Raw onions
- Citrus fruit or peels
- Chocolate
- Moldy or rotting food (soft or undesirable for humans is OK, moldy is not)
- Dry or undercooked beans (fully cooked or canned beans are OK)
- Styrofoam*

* I would never expect that you might consider intentionally feeding your chickens Styrofoam. Just be sure not to leave a Styrofoam plate or takeout container where they can get to it because, despite it being horrible for them, chickens seem to love eating it!

41
Fun with Fodder

Fodder is an affordable way to bring fresh greens to your chickens even during the winter! Fodder is made by simply allowing seeds to sprout and green up. It's a great treat for run-bound chickens who don't have access to fresh grass. Pound-for-pound, fodder will cost less to feed than even fermented food. To grow enough fodder to feed your chickens full time is quite space-intensive, but even a small backyard flock can benefit from the fodder being part of their snack rotation by following the procedure below.

How to grow fodder

Supplies

- 1½ Tbsp organic wheat or rye berries
- Quart jar
- 1 tsp vinegar
- Large cake pan or plastic container
- Sunny window

Directions

1) To a quart-sized jar, add the berries and enough cold water to the cover the seeds. Swirl and pour off the dirty water.
2) Refill the jar with water. Add 1 teaspoon vinegar and leave to soak for 10 to 12 hours or overnight before draining.
3) Rinse and drain the seeds every 8 to 12 hours.
4) On the second day, you will see tiny tails growing from the seeds. Dump them out into a cake pan, shaking it to evenly spread the seed.
5) Continue gently rinsing the seeds and immediately draining the cake pan 1 to 2 times a day. After a few days, they will develop an interlocked root system that will allow you to lift the mat up by the grass without it falling apart.
6) On day five, move the fodder near a sunny window so it can absorb more sunlight and increase the nutrient content through photosynthesis.
7) Remove the fodder mat from the pan and feed it to the chickens on or about day seven or when it is as tall as you would like it to be.

42
How to Make Grazing Frames

Less ongoing work than growing fodder (page 85), grazing frames are another way to allow your chickens access to greens. The frame protects the bottom portion and root system of the grass, which keeps your chickens from decimating the area and turning it to a mud pit.

How to build a 4 × 4 grazing frame

Supplies

- 8-foot-long 2 × 4 wood (you need 2)
- Drill with screwdriver bit
- 2.5-inch wood screws (you need 8)
- 20 wood fencing staples
- 4 × 4 piece of ½-inch hardware cloth
- Hammer
- Wheat berries, rye berries, or other chicken-friendly plant seeds

Directions

1) Cut the 2 × 4s in half so that you have four, four-foot sections. If you don't have a saw at home, most hardware stores will cut your 2 × 4 in half for you if you ask!
2) Lay the wood pieces out on the ground in a square.
3) Use the drill to drill two pilot holes at each corner. Switch to the screwdriver bit and drive two screws at each corner to hold the frame together.
4) Lay the hardware cloth over the frame and secure it along the edges by hammering in fencing staples.
5) Set the frame, hardware-cloth side up, in your desired location. Sprinkle wheat or rye berries or grass seed through the hardware cloth and water well.

If you set your frame in a sunny spot, in about seven to fourteen days (depending on temperature outside), you should start to see bits of grass growing up above the frames! The chickens will be able to nip the tops off as they grow through without damaging the root system underneath.

43
Water Additives

If you've ever enjoyed a refreshing fruit-infused water, you know how lovely water additives can be. For chickens, some water additives can provide health benefits along with an interesting change of flavor.

- **Electrolytes:** Add this powder to water for freshly hatched chicks or any time a chicken is stressed or injured. Follow the instructions on the package for proper dilution ratio.
- **Apple cider vinegar:** Has probiotic and antibacterial properties. Apple cider vinegar improves the flavor of the water to the chickens and it encourages them to drink. Be sure to choose the apple cider vinegar *with mother*. Add 1 tablespoon per gallon of water.
- **Fresh herbs:** You can add fresh herb trimmings to your chickens' water. Some flavorful suggestions include mint, basil, rosemary, and lavender. If you serve this fancy water treat in an open-top bowl instead of their regular waterer, they can enjoy nibbling at the herbs, as well.
- **Dried herbs:** You can repurpose the same herbs you used to make your nesting box spa blend (see page 59) to make a flavorful herbal tea for your chickens. Steep a tablespoon of the nesting box blend in a cup of hot water and add it to a gallon of fresh water.
- **Molasses:** By itself or as part of a proprietary product such as Nutri-Drench™, molasses can be useful as part of your chicken first-aid kit to help sick or injured hens. Mix ¼ to ⅓ cup (measurement does not have to be exact) to about one gallon of water. Even though chickens like the flavor of molasses, save this for when they are ill and don't use it all the time—too much molasses can cause diarrhea.

44
Make a Snack Block for Your Flock

A Flock Block is a popular proprietary product made by Purina and is the most popular chicken snack block on the market. It is 25 pounds of seeds and grains held together in a block form. It's a great way to keep chickens occupied and pecking at things that aren't each other.

While Flock Blocks are pretty affordable, the grains in the block are likely GMOs (see page 77). Since we try to feed organic, I like making my own so I am in charge of the ingredients. This block shouldn't be used as your chickens' only source of food but makes a great treat or *snacktivity*.

Ingredients

- ¾ cup blackstrap molasses
- ½ cup melted coconut oil, lard, or bacon grease
- 4 eggs
- 2 cups scratch grains or wild bird food
- 2 cups whole grain chicken feed
- 1 cup flax seeds
- ½ cup rolled oats
- ½ cup chia seeds
- ½ cup flour*

*Whole wheat flour or "cup-for-cup" style gluten-free flour works for this.

Directions

1) Preheat an oven to 350°F and prepare a loaf pan by lining it with parchment paper.
2) In a medium bowl, whisk together the molasses, fat, and 4 eggs (save the shells!).
3) In a large bowl, mix the remaining ingredients and the crushed shells from the eggs used in the previous step.
4) Add the wet ingredients to the dry ingredients and mix until well combined.
5) Press the mixture into the prepared loaf pan and bake at 350°F for 45 minutes.
6) Remove the snack block from the oven and allow it to cool completely while in the loaf pan (it will crumble if you remove it too soon!).
7) Set outside for your flock!

45
Make an Edible Piñata

Having a variety of tasty activities in your arsenal as a chicken keeper is a good way to keep everybody happy and healthy. Chickens love eating all kinds of fruits, veggies, and grains. If you can keep them interested in their snacks because you hung them up, they will take more time to eat and stay interested longer.

It's easy to turn a variety of foods into a piñata-style treat. Try poking a hole in one of these foods and threading twine or a hook through to hang it in their run:

- Apple
- Basil (tie as a bundle and hang upside down)
- Cabbage
- Carrot greens (bundle together and hang upside down)
- Corn on the cob
- Cucumber
- Kale (tie as a bundle and hang upside down)
- Mint (tie as a bundle and hang upside down)
- Melon
- Pumpkin

I find that if my chickens haven't started eating something after a couple hours, cutting a slice off to expose the inside encourages them to taste it.

You can also make a basket-style piñata by using an empty wire-hanging basket frame (the kind designed to be used with a coir lining) and filling it with salad greens, herb trimmings, or kale stalks and hanging it at chicken level in your run.

46
Start a Mealworm Farm

Mealworms are the larvae of the darkling beetle, which are considered a delicacy by chickens. You can purchase freeze-dried mealworms from the feed store, but they can be pretty pricey if you give them as more than an occasional treat. Growing your own at home isn't hard, and chickens seem to love them even more than the freeze-dried variety! Raising your own mealworms is a fun project to do with your kids that will help them learn more about the darkling beetle life cycle.

How to raise mealworms
Follow these simple steps to grow a colony of mealworms to feed your loyal subjects, er, chickens.

1. Find a clear 5- to 10-gallon container to use as your mealworm farm. An empty aquarium or plastic tote makes a good container.
2. Use a screen to cover the container so the airflow is not restricted. This type of cover is designed to keep things out of the mealworm farm rather than keep the mealworms in.
3. Put 3 to 4 inches of either wheat bran or rolled oats in the bottom of your container to serve as bedding. Add a fresh carrot or a potato cut in half for food.
4. Add the mealworms into the container. A good rule of thumb is to start with about 100 meal worms per every gallon your container holds (so 500 meal worms for a 5-gallon container).

How to care for your mealworms
Mealworms are easy to raise; however, they are sensitive to moisture. The container must be kept in a dry area indoors like a garage or laundry room. If the container is put into direct heat, rain, or cold, the mealworms will die.

Feed your mealworms root vegetable matter like fresh potatoes or carrots. As the wheat bran or oats get low, be sure to top that off, as well.

Don't be alarmed if you see mealworms start to turn into beetles. The lid will keep them in the container and the eggs they lay will turn into more mealworms. You can pull out dead beetles to feed to your chickens or leave them in the mealworm farm and the new worms will eat them.

47
Herbs for Chickens

Herbs have lots of wonderful health benefits for your chickens. Just like people, chickens can enjoy herbs lots of different ways!

Fresh herbs

Chickens love fresh herbs! From mint, basil, rosemary, thyme, parsley, and more, chickens love fresh herbs served these ways:

- Growing in their run for fresh snacking (see page 154)
- Fresh herb trimmings to scratch through and peck at
- In their water to make a flavorful, hydrating treat
- Bunches hung in the coop to deter pests (and for snacking)
- In their nesting boxes

Dry herbs

While not as nice as fresh, chickens enjoy dried herbs, too! Try one of these ways to treat your hens to a dry herb-based treat:

- Steeped as tea (see page 89)
- Nesting box herbal scatter (see page 59)
- Mixed into their food
- Sprinkled over fresh coop litter
- Mixed into a snack block (see page 91 for recipe)

48
Get a Boost

Chickens get nutrients through their feed, snacks, and range time, but boosting their nutrients through supplements can help make sure they are getting everything they need. Different types of supplements can help chickens in certain situations boost their immune systems and even appetites.

Types of supplements

- **Calcium:** Laying hens use lots of calcium in the process of making eggs. Make sure your hens have enough by adding crushed egg shells or oyster shells to their run for them to peck at.

- **Diatomaceous Earth (DE):** DE is considered a natural wormer, and in one study hens whose feed was 2 percent DE laid more and larger eggs than the control group.[3] Look for DE labeled as "food grade" and mix it into your dry or fermented chicken feed.

- **Electrolytes:** These are important to have on hand to boost chicken health and hydration. I like to use electrolytes in the first batch of water I give to chicks and keep it on hand in my chicken first-aid kit to help ill or stressed hens.

- **Grit:** Since chickens don't have teeth, they eat small rocks called grit to help them break down their food. Grit can be added to your chicken feed or given in a free choice container.

- **Herbs:** The same herbs used to make a nesting box spa blend (page 59) make a nice dietary supplement for chickens. Mix it into their feed or just allow them to pick through the herbs when they visit the nesting box to lay their eggs.

- **Probiotics:** Probiotics can come in powder or liquid form. These are used when chickens need to improve digestion, appetite, or immune system. They can be useful for getting a hen's digestion back in balance after a course of antibiotics.

---

3   Bennet, Yee, Rhee, and Chang. "Effect of diatomaceous earth on parasite load, egg production, and egg quality of free-range organic laying hens." *Poultry Science*, July 2011. https://www.ncbi.nlm.nih.gov/pubmed/21673156

Chicken Quirks

It's something you hear but don't believe until you become a chicken owner: chickens have a lot of personality! In addition to the personality and quirks of individual chickens within the flock (yes, really!), chickens as a whole have some behaviors we find quite amusing. Learn more about some of the quirky things chickens do and how you can support them in their peculiar yet totally normal behaviors.

49
Are Chickens Scared of the Dark?

While chickens have slept in the dark for as long as they have been on Earth, you wouldn't know it based on the way some of them behave! When my first flock of chickens was just a few months old, I had a hard time getting them to go into their coop at night because they wanted to huddle by my back door near the light that was coming through the glass. Chicken forums online are likewise full of stories about chickens who will cry at their owners and make quite a fuss if they have a light on in their coop and it is suddenly turned off. So what gives?

Maybe "scared of the dark" is the wrong way to put it, but "preference to be in the light" is right. Chickens like light, which makes sense because light has some distinct safety advantages.

How to help your chickens in the dark

- If your chickens don't want to come into a dark coop at night, a nightlight powered via batteries or an extension cord will help draw them in.
- Chickens can be gradually trained to tolerate the dark. Start with a low brightness bulb in the coop, gradually swapping it out for dimmer and dimmer lights until they are more accustomed to the dark.
- The "rip off the bandage" approach is to shut them in their coop at night and just switch of the light despite their protests. While not the most compassionate approach, I am not aware of any chickens who have died due to fear of the dark.

50
Bathing in Dirt

As funny as it sounds, chickens actually take baths in dirt to get themselves clean. To see a hen dustbathing is quite the sight! They get low to the ground and use their feet to kick up dirt onto their backs and against their skin at the base of their feathers. While you can let your chickens dig a hole to make a dust bath any old place (and let's face it: they will!), making them a dedicated dust bath area full of beneficial dust-bath-boosters ensures they always have what they need to stay clean and parasite-free.

How to make a dust bath for your chickens
Choose a dry area either inside the coop or in another covered area.

- Dig a wide hole or use a container like a shallow tub or even a kid-sized swimming pool. If you use a container with a bottom, drill some holes for drainage in case rain gets inside.
- Fill your container or hole with loose dirt. You can use dirt from your yard if it has low lead levels (try mixing it with sand to make it looser as needed). You can also use pre-packaged potting soil or sand. See page 71 for more information on testing your soil for lead to help keep your family's toxin exposure down.
- Add some mix-ins to boost your dust-bath power:
  - Food-grade diatomaceous earth kills bugs like fleas, ticks, and lice.
  - Wood ash contains beneficial minerals and also repels parasites in a similar way to diatomaceous earth.
  - Dry herbs smell nice and help repel undesirable creepy crawlies. You can use the same herbs from the nesting box herb blend (page 59) for this purpose.

51
Teach Your Chicken Tricks

Despite insulting terms like *bird brain,* meaning you are a bit dim, chickens are really quite clever. Many people have successfully trained chickens to perform tasks and do tricks. Chickens can be taught to run agility courses, play musical instruments, play tic-tac-toe, and more.

Teach your chickens to come when called

This might be the easiest thing to teach your chickens to do. Decide what you want to use as the signal that brings them running; a verbal command, clicking your tongue, or a noise such as shaking coins in a can are common. Every time you go out to feed treats, say the command or make the noise right before and during feeding the treats. The chickens will learn to run your way when they hear that noise in anticipation of a reward. In addition to being amusing, this is a great skill for them to have if you ever need to relocate them or to get them into their coop for early lockup.

You can also train individual chickens to come when you call them by name by having individual training sessions with that chicken separately. Use the same approach as above but say the chicken's name as you give it treats. It's important to do this one-on-one because if the other chickens are around during training and see Elsa getting treats every time you say her name, they will learn to run to you when you say "Elsa"—even if that isn't their name!

Teaching tricks

The trick to teaching your chickens to do tricks is to motivate them with treats. Use something they *really* like eating, such as mealworms or soldier fly larvae. Also, keep your training sessions short. You'll be surprised how quickly chickens pick up on things and how much you can accomplish in just a few minutes a day.

Teach chickens simple one-step tasks and tricks like jumping straight up or hopping off things the same way I described teaching them to come when called, above. For more complicated tricks, work in incremental steps leading up to getting them to actually do what you want.

For example, you can teach a chicken to jump on your lap on command by first giving it treats for coming near you while you are sitting on the ground. Then for touching you while you are on the ground. Then for climbing onto your lap. Then coming onto your lap when you move to a chair. Once you have the chicken doing what you want, introduce the command and continue rewarding when the chicken does what you'd like.

Just like with dogs, it is important to run through the commands and tricks your chickens know from time to time so they don't forget the commands and need to be re-taught.

52
Baby Talk

Baby chickens have a language all their own. You'll hear a collection of different peeping patterns coming from your brooder box. As they grow, you will start to hear different "vocabulary" coming from them as they transition from being babies to grown-ups. In the meantime, here is some insight into what some of these chick noises mean:

- **Happy chirps:** The happy, social chitter-chatter between babies as they talk and interact with each other. This is the noise you will hear most often from happy, well-cared-for chicks.
- **"Something's happening":** When something interesting happens like a hand reaching into the brooder or a chick discovering a treat, chicks will cheep at a faster pace and also run around.
- **Distress chirps:** A high pitched, loud "cheep! cheep! cheep!" on repeat is a call to their mother (if you are raising them in a brooder, that's you!) that something is wrong. Maybe their water dish fell over. Maybe they got stuck. Maybe they're lonely. Maybe they are cold. Better go check it out.
- **Sleepy noises:** This sounds like a soft cooing or purring noise. You'll hear happy babies make this noise when they get comfortable and settle in for the night. If you are lucky and they feel bonded to you, they may also make this noise if they settle in to take a nap in your hand.

53
I Shall Sing You the Song of My People

While most people are familiar with a rooster crow, hens all seem to have their own common chitter chatter as well. From the egg song to their distress call, check out what some of those noises coming from your yard mean:

- **Greeting clucks**: Sounding similar to "Brr-gup," chickens make this noise as they greet and interact with each other. This is a bit like the more sophisticated adult hen version of "happy chirps" from page 109.
- **Check this out:** Low noises like "buk-buk-buk bu-bawwwwk" is how chickens tell each other to come check stuff out. Maybe they discovered a tasty treat or found a cool new toy.
- **Alarm noises**: If one or more of your hens starts making a sound that reminds you of a fire alarm, drop everything and go see what's wrong. This is the sound they make to warn each other when they spot predators, and your presence may be needed to help protect them or to scare off what spooked them.
- **The egg song:** Chickens like to announce that they have laid an egg with a distinct song that sounds a bit like "brr brr br-kawk!" Sometime hens get so excited for their flock mates to lay eggs that they will sing the egg song for each other almost as if to say "congratulations!"
- **Crowing:** When juvenile roosters learn to crow, watching it happen is quite a sight! It's almost as if you can see a crow rolling up their throat to practically be vomited out of their mouth. If this happens, you'll need to decide if you want to raise a rooster. If not, head to page 209 to decide what to do with him. If a full-grown adult hen starts crowing, however, that's a different story entirely. In flocks without roosters, the lead hen will sometimes take on the rooster role—complete with crowing! Sometimes she will also stop laying eggs while assuming this position. In my flock, this has happened occasionally but been short-lived. Some hens assume the rooster role permanently, however.
- **Growling:** If you find your hen spread out on a nest and growling at you or any other chickens that come near her, you probably have a broody hen on your hands. Turn to page 129 to learn what to do about that!

54
Start a Chicken Rock Band

Millions watched Jokgu the chicken play "America the Beautiful" on the keyboard during season 12 of America's Got Talent—and millions more have seen the replays online! While Jogku and her sister Aichan were eliminated after only two appearances, they proved to the world that chickens really can rock.

You can start your own chicken rock band at home, too! Try these fun chicken rock band ideas to get started:

- **Preschool piano:** Get a toddler or preschooler sized piano. Try putting meal worms or another favorite treat on the keys to encourage the chickens to peck. You can also do a web search for "teach chicken to play piano" for some interesting videos showing how to do it!
- **Xylophone boredom buster:** Buy a colorful child's xylophone and hang it on the side of the coop. Try drawing a single dot in the middle of each bar to encourage them to peck where it will make more noise.
- **Drums:** Pick up a desktop drum set or a toddler toy drum. Draw a dot in the middle with a sharpie to encourage pecking and encourage your chicken along with treats.
- **Light up keyboard:** Try encouraging the chicken to peck the keys by using a keyboard with lighted dots under the keys designed to show you which keys to press to play a song.

No matter which instrument you use, positive encouragement and treats are the way to a chicken's heart. If the chicken looks stressed at any time, let it go and come back and try a different time.

55
Skip the Sweater

I am a knitter. In the twelve years I have been a knitter, I have seen people knit clothes and cozies for just about everything from teapots to dog to apples. One thing you shouldn't knit a sweater for, though, is your chickens.

Just say no to sweaters

Unless your chicken has completely bare patches of skin because she was rescued from a battery hen operation or is going through a very bad molt, sweaters will do more harm than good. The way chickens stay warm is by fluffing themselves up. When they do this, their feathers trap tiny air pockets of warm air near their skin. This is how they stay warm all night during the winter.

What happens when you put a sweater on your chicken? It can't fluff! The feathers are held down and the chicken is no longer able to fluff itself to stay warm. So even though chickens in sweaters are adorable, just say no to the sweater.

Vests are okay

Also known as "chicken saddles" or "chicken aprons," these vest-style fabric devices are designed to protect chickens from roosters who are overly aggressive lovers. If you are starting to see wear on your hen's back with cuts or feathers missing due to mating, a chicken apron is a good way to protect her. Unlike sweaters, the woven fabric design both protects the hens from the rooster's spurs and is stiff enough to lift off the feathers for air flow underneath. This both allows her to fluff and protects her skin until the feathers grow back underneath.

If you feel like your hen needs a chicken apron, you can buy one at feed stores or online. If you are crafty, you can find patterns for chicken vests or chicken aprons online, as well. Just be sure to use a woven fabric instead of a knit, as stiffer fabric will make it easier for your chicken to fluff while wearing it.

56
The Pecking Order Is Like High School

While the pecking order does involve pecking, it isn't so much about who is allowed to peck as it is about who is the lead or queen hen. Chickens have all kinds of behaviors to express dominance including puffing up feathers, flapping, jumping on each other, pecking, squawking, and more. They will sometimes peck at each other, but this doesn't have to be as aggressive as it sounds, as pecking and nudging each other with their beaks and heads is just one way chickens communicate with each other.

Deciding who is in charge starts in the first few weeks. In a brooder, chicks will sort out among each other who is lead chick. In my flock, my original lead chick is still my queen hen (Elsa, the chicken on the cover).

If a broody hen is raising the babies, they all start at the bottom of the hierarchy by default and sort out their place in the flock as soon as they start interacting more with the grown hens. Grown chickens will shoo the tiny ones away from food and take the best roosting spots (and boot lower-ranking chicks off the good roost if one of them got there first).

Despite the myth that the pecking order is all about "who is allowed to peck who," the pecking order really seems to be more about who gets the best stuff. Who gets the best spot on the roost? Who gets to kick another hen out of the best nesting box? Who eats first? Who gets the best snacks? The queen does. Sometimes deciding who is queen involves a fight, but not often.

Just like high school, chickens can also form cliques or smaller groups of hens who tend to hang out together. Also like high school, the higher-ranking flock members usually hang out together while the lower-ranking flock members form their own group. While you may still see the occasional squabble and a higher-ranking chicken putting another in its place, part of the purpose of the pecking order is to make sure every chicken knows where they rank so they don't have to fight about it all the time.

57
Easter Egg Hunting

Even though the idea is generally that chickens should be laying their eggs in the nesting boxes, sometimes chickens get other ideas. Has it been a while since your chickens have left you eggs? Are you only getting half as many eggs as you expected? It could be because of natural circadian rhythms or it could be because one or more chickens has created a secret nest you don't know about!

This, of course, is a bigger problem if you have a rooster. A rooster means that any hidden eggs could possibly be fertilized, and if a hen is off secretly sitting on them, it might be an even bigger problem.

If you have reason to suspect your chickens have a secret nest, it's a good idea to walk your property to see if you can find it. Look behind things and under things; hens like cozy little hidden places to lay eggs so you might find a nest behind some boards or under a bush.

If you find eggs on your little hunt, your next step will be to see if they are any good for eating. If you have a rooster, follow the instructions on page 23 to candle the eggs and make sure baby chicks haven't started developing inside of them. If you don't have a rooster, the float test will be sufficient to figure out if your eggs should still be good for eating. You can find instructions for float-testing eggs on page 171.

58

Fairy Eggs, Monster Eggs, and What the Heck IS That?

If you have always bought your eggs from a grocery store, you may not know that not all eggs are the same. Occasionally something funny happens to your little egg factories that leaves you laughing, scratching your head, or both.

Fairy eggs (sometimes called rooster eggs, fart eggs, or wind eggs)

What it is: An extra tiny egg without a yolk

 What causes it: The yolk wasn't ready before the chicken started creating the eggshell. This is most common in new layers as well as in older hens who might be approaching "hen-opause."

 What to do about it: Marvel at how funny nature is and maybe post a photo on Facebook. Fairy eggs are nothing to be concerned about.

Double-yolked eggs

What it is: An egg containing two yolks. These are often larger than the ones regularly laid by that hen.

 What causes it: An extra yolk was released during ovulation and caused two yolks to be bound in the same shell. These are more common in young hens just beginning to lay.

 What to do about it: Feel free to eat it and consider your good fortune— double-yolked eggs are often considered good luck! Don't try to hatch these eggs, though. Due to crowding and placement of the air pocket, double-yolked eggs rarely hatch successfully.

Egg-within-an-egg

What it is: A monster egg that, once cracked open, reveals a yolk and white as well as another smaller egg.

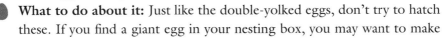 **What causes it:** A fully-formed egg backed up into the reproductive tract instead of continuing toward the vent, causing it to become encapsulated in the egg forming behind it.

What to do about it: Just like the double-yolked eggs, don't try to hatch these. If you find a giant egg in your nesting box, you may want to make a video while cracking it open to share on social media or with your other chicken-loving friends.

Egg without a shell

What it is: A soft, rubbery egg with a yolk and white encased in a membrane but not a shell.

 What causes it: The hen's shell gland was not in sync with the rest of her reproductive tract, or your hen may be deficient in a nutrient like calcium needed to form eggshells.

 What to do about it: If your hens are eating layer feed and one of the first eggs comes out without a shell, it may just be that her shell gland is still not in sync with the rest of her system. If a chicken lays shell-less eggs regularly or multiple times in a row, make sure they have adequate access to a calcium supplement like oyster shell or even dried and crushed eggshells.

Lash egg

What it is: Unlike a soft egg without a shell, a lash egg is not an egg at all, but mass of puss and other material that traveled through the oviduct that may resemble an egg once laid.

 What causes it: Lash eggs are caused by an infection called Salpingitis.

What to do about it: Salpingitis can be fatal to hens but one lash egg does not a death sentence make. You may want to consult a vet about administering antibiotics to the affected chicken.

59
Chicken Run Boredom Busters

If you have your chickens confined all or most of the time instead of ranging, boredom is an issue you will need to deal with. If the chickens get bored, they could start pecking at each other for the sake of some excitement—and someone might get hurt real fast! If your chickens stay confined, make sure you provide ways for them to be entertained before injuries become an additional problem to solve.

Not sure what to do to bust through flock boredom? Try one or more of the suggestions below:

- Make a piñata from a round chicken-friendly fruit or vegetable like an apple or a cabbage. See page 93 for some ideas on what to hang.
- Leave musical instruments accessible in the run. See page 113 for more ideas and suggestions about chicken-friendly instruments.
- Get a flock block to serve as a "snacktivity" alongside their regular feed. Get directions for how to make your own version on page 91.
- Try installing a couple chicken swings. You can buy commercial swings or make your own with a small log and rope.
- Put snacks in a treat ball. It's fun to watch your flock race around the run kicking the ball as worms fall out—almost like a chicken soccer game.
- Give them a pile of trimmings to scratch at and pick through. Straw, hay, yard trimmings, or fallen leaves make great pile materials. They will scratch through until they have found every last bug and seed and spread it out in a nice layer for you.
- Install extra roosts in their run area so they have nice places to sit at a different height and observe what's happening.

60
How to Tell Chickens Apart

My chickens are pets that have the side perk of providing my family with breakfast. They each have a name and I wanted to be able to tell them apart from each other. That's one reason I chose a variety of different chicken breeds—because if they all look different, it's easy to know which chicken I am looking at.

Not everyone keeps chickens as pets and many people don't name any except their most favorite chickens. If you want to be able to tell your chickens apart but they look too much alike, try one of these tips:

- **Poultry ID bands:** These bands come in many different styles and colors. Some even come with fun charms or symbols on them. Choose a different band for each chicken and all you'll have to do is look at her feet to see which chicken is which.

- **Zip ties**: Can't find poultry ID bands? Different-colored zip ties work the same way in a pinch. Be sure not to put the band on tight so it doesn't cut off circulation (also make sure it's not so loose that it falls off!).

- **Poultry pedicure**: We have a family friend whose daughter once gave her favorite chicken a pedicure to match her own. If you have different colors of nontoxic nail polish, some patience, and a helper, you could paint the nails of each chicken a different color.

- **Color spot:** Use a nontoxic food coloring or dye and put a different-colored dot on the top of each of their heads or another location like their chests, backs, or shoulders. This has the potential to draw the attention of the other chickens, so be sure to observe carefully if you choose to do this to make sure nobody is being pecked too much. I also recommend using colors other than red.

61
How to Help Your Ladies Through Molting

Although it might feel alarming the first time you start seeing chicken feathers all over the run, molting is the very normal process by which chickens grow fresh feathers. As new feathers grow in, they push the old ones out and you will find them on the ground. During this process, chickens usually stop laying and divert all their nutritional resources to growing feathers. Ideally, a chicken won't have any bald spots while molting, but during a "hard molt" this can happen if the chicken is particularly stressed.

To help your chickens get through molting in the quickest and most stress-free way possible, turn off their supplemental winter lighting (if you are using it, see page 201) and feed them some extra protein.

Great sources of protein for your chickens include:

- Meal worms (fresh or dried)
- Soldier fly larvae (fresh or dried)
- Scrambled eggs
- Meat scraps
- Broiler chicken feed
- Protein powder mixed into their fermented feed (see page 79)
- Unsalted nuts
- Crickets
- Fresh worms from the garden

My kids like to help their chickens by going on bug and worm hunts through the yard and garden and delivering these juicy morsels to their favorite birds.

While adding protein to their diet during a molt will help support them, be sure to keep feeding them their regular layer feed at the same time (unless you temporarily switched to broiler feed for the higher protein content) to help meet their other nutritional needs.

62
What to Do with a Broody Bird

A broody bird can be a blessing or a curse, depending on what your goals are for your flock. If you're all about eggs, a broody bird isn't helping you meet your goal! If you were hoping to expand your flock size, though, you are in luck!

If you realize one of your chickens is brooding, it's important to take one of the actions below. Allowing a hen to brood indefinitely can cause major problems for their health, as a dedicated broody will sit for weeks at a time hoping to become a mom. If you aren't going to let her raise at least a baby or two, it's important to break her so her health does not suffer.

Broody option 1: Let her raise babies

Broody hens are wonderful at their jobs and have a work ethic like no other. A broody hen will only leave the nest once or twice a day to poop, eat, and drink. Using a broody mom to raise chicks for me is my favorite way to expand my flock. For more information on what to do if you want a broody to raise babies for you, go to page 19.

Broody option 2: Break her

The process of getting a chicken to stop being broody is called "breaking" her. If you don't want to use your broody hen to raise babies for you, you'll need to break her of being broody. It's simple to do but requires consistency.

- Pick up your broody hen off the nest and lock her outside of the coop so she can't sit on the nest for the rest of the day.
- Promptly remove all eggs from nesting boxes after they're laid to prevent her from sitting on them.
- Block access to nesting boxes at night to encourage her to sleep on the roosts.

Sometimes these things alone are enough to convince her she didn't really want to be a mom anyway. If, after a few days of removing her from the nest, she keeps going back or appears desperate to nest, place her in a crate with a wire floor (a rabbit hutch is perfect!) that is elevated off the ground so there is air flow under her bottom. This helps prevent her from maintaining the temperature she needs to feel the urge to brood. Keeping the area well lit further helps break her of her broody tendencies, as snuggly, dark places encourage her urge to brood.

Keep her in this crate (with food and water, of course) for two or three days before letting her return to the flock. If, after a few days in jail, she tries to go right back to brooding on the nest, you can put her back in broody jail for another couple days. Alternatively, if it is a nice, warm day, you can try giving her a cool bath and setting her outside to scratch around while she dries. Continue alternating days in broody jail and cool baths as needed until she resumes her normal chicken activities.

An Ounce of Prevention—Avoiding Things that Could Go Wrong

🐔 🐔 🐔

Just like with most things, it is easier and less costly to prevent problems with your chickens now than it is figure out how to fix problems later. From predator-proofing to having essentials on hand "just in case," these tips and tricks will have you prepared for Murphy before he even gets his shoes on.

63
Skirting the Issue

One of the biggest threats to your chickens in the evening when they are sleeping is ground predators. Foxes, dogs, coyotes, and more will come in to your yard and test the limits of your predator-proof chicken buffet. If your coop is closed up tight, one way a predator might try to get into your coop is by digging. Adding a skirt around the perimeter of your coop and/or run will help keep predators from being able to dig down to get in and snack on your flock.

Bricks or pavers
Some people create a surface-level skirt of sorts by laying bricks or pavers around the outside of the perimeter of the coop and/or run. The thought behind this is that an animal might walk up the edge of the fence but be unable to dig to get below it because a dig-proof hard surface is there.

Pros
- Less digging required to install this versus a wire skirt.
- Uses inexpensive materials.

Cons
- Width of border has to be enough that the animal wouldn't know to just go dig on the other side of the paver, so it needs to be multiple bricks or pavers thick—which requires more work to set properly.
- Materials are heavy to move.

Wire skirts
The premise of the wire skirt is that you bend a durable wire material such as hardware cloth or welded wire at a 90-degree angle to create an L shape. The vertical leg of the L goes against the side of the coop while the horizontal leg extends out. While some people lay the L on top of the ground and bury it under only an inch or two of dirt, the most secure way is to dig down 18 to 24 inches, extending the wire down that length before turning it at 90 degrees for another 18 to 24 inches out from the wall.

Pros	Cons
● Very secure.	● Requires a lot of labor-intensive digging.
● Materials weigh less for easier transportation to coop site.	

The right solution for you likely depends on the kinds of predators you have in the area so consult with some other local chicken keepers to help you decide which is more appropriate.

64
Set Up a Light-Based Security System

While a strong, predator-proof coop is essential, a light-based security system can help keep your birds safe by deterring predators before they even get close enough to your coop to test its strength. The lights play off of nocturnal animals' fear of being seen. One system startles them away while the other gives the impression they are being watched by another animal.

Motion-detecting light systems

This type of light system is designed to startle away animals lurking around by turning on a bright light when it detects motion. Many motion-detecting lights are now available that are either solar- or battery-powered (or both!), so you don't have to worry about running electricity to your coop or paying an electrician to install a hardwired light.

If you use a system like this, be sure to play with the placement of the motion detector so it can detect small animals close to the ground like raccoons or bobcats. You might find that installing it just a few feet up on the wall of the coop works better than installing it near the roofline.

Eye-simulating light systems

These types of systems have you install a small unit that begins blinking a red light at dusk. Most of these systems also run on either solar or batteries to make their placement very flexible.

The blinking red light is designed to mimic the way an animal might perceive light catching on eyes (the same phenomenon that causes "red eye" in flash photography) of a larger predator. If the animal considering your yard as the source of its next meal thinks it might be detected and possibly attacked by a larger predator, they may avoid the area entirely.

65
How to Choose Raccoon-Resistant Coop Locks

Unlike dogs, foxes, and coyotes, raccoons pose a unique challenge as far as ground predators go because they have hand-like paws and are smart. The general rule of thumb is that if a four-year-old human can open it, a raccoon probably can, too. Anyone who has had a four-year-old child knows: four-year-olds are masters at getting into things! Raccoons are pretty strong, though, so also be thinking "If a four-year-old were as strong as I am, could he do this?"

Components of raccoon-resistant systems

Anything that uses a simple mechanism to open—turning a handle, lifting a door, flipping a latch—is easy enough for a raccoon to do. Anything that requires multiple actions simultaneously, such as holding a twist lock open while also pushing in a lever like some locking carabiners have, is more appropriate.

It's also important to make sure other potential areas of weakness in your coop are addressed. Your raccoon-proof lock won't do you any good if the raccoon gives up on your latch and decides to peel back the chicken wire covering the windows instead.

Raccoon-resistant lock suggestions

Here are some suggestions for lock types that many people find work well at keeping raccoons out of their coops:

- **Locking carabiners** that require you to twist, squeeze, or unhook two different latches at the same time to open them.
- **A key-operated lock** through a metal latch. Some people leave the key outside, attached to their coop with a string or small chain, so it can't get lost, while others keep the key inside with the humans.
- **A combination lock** through a metal latch. I suggest a locker-style spinning-number lock, as I have seen reports online of determined raccoons returning night after night to eventually decode the simpler combination locks where you line up letters in a row before it unclasps.
- **A key-operated deadbolt** similar to the type used for front doors on homes. If you do this, be sure to keep the spare key somewhere safe, because if you lose it, you can't just cut off the lock with bolt cutters like you can with the other lock types.

66
Shoo, Hawk!

The main overhead predator most chicken keepers in North America are worried about is hawks. Hawks mainly prey on small to medium–sized animals like tiny birds, mice, rats, and squirrels—so your chicks are especially susceptible when outside. While smaller birds are the most vulnerable because they are easy to catch without a fight, a hungry hawk will still go after full-sized chickens, so be sure to use methods that defend all areas where your chickens hang out.

> **IMPORTANT!**
>
> It is illegal in all parts of North America to harm, harass, or kill any species or age of hawk. Hawks are protected under the Migratory Bird Treaty Act. Your hawk-deterrent strategy should focus on making your yard less appealing to them but may not involve methods that harass, harm, or kill hawks or their chicks.

To help make your yard less appealing to hawks, try one or more of these strategies:

- **Hang old CDs:** The reflective side of the CDs is thought to reflect enough light to confuse the birds and make the areas where they are hanging unappealing.
- **Scare tape:** This reflective ribbon is commonly used to repel wild birds from farms and gardens. You should be able to pick it up at any hardware store or garden center. Even if you have an open-top run that's too large to practically cover with hardware cloth, consider stretching some scare tape across the top to keep predatory birds from using your run as a lunch box.
- **Get a guard dog:** Some dogs, especially Livestock Guardian Dogs (LSDs), have a calm demeanor and strong drive to protect. Birds don't usually want to mess with large dogs, so just having your large chicken-friendly dog around is a good deterrent for anything that is considering snacking on your chickens.
- **Use a scarecrow:** Only the most desperate of hawks would try to steal a bird with a human standing right there. Of course, you can't be there all

the time, which is where the scarecrow comes in. Be sure to move him to a new spot every day so any predatory birds casing your property don't figure out that "you" only ever stand in one spot.

- **Get an owl:** A real owl is a threat to your chickens, but some people have had success with a fake decoy owl spooking hawks away. Just like the scarecrow, plan to move your owl to a new spot occasionally.
- **Fishing line:** Use fishing line to run back and forth across the top of an open run. If you have a large yard, you could put up tall posts on the edges to run web between. The light reflects on the fishing line, making hawks unsure if the area they are scoping out has a roof. Since they want to be able to get in and out quickly, a makeshift web like this might be enough to get them to move on to other things.

67
How to Build a Chicken First-Aid Kit

Occasionally a first-aid situation will occur with your flock and you'll need to have some essentials on hand. In the middle of the emergency is NOT the time to make a feed store run to get what you need! Keeping a chicken first-aid kit handy so you can care for them properly will help you take care of issues in a timely matter, which can make a big difference in keeping little situations from turning into big ones.

First-aid kit essentials

These are the minimum essentials needed for a poultry first-aid kit. I like to store my supplies in a waterproof plastic box.

- **Electrolytes:** Powdered electrolytes can be found at your local feed store. To use, simply mix it into your chickens' water following the instructions on the package. Electrolytes help when chickens are stressed or dehydrated.
- **Gauze and vet wrap:** These are great to cover abrasions on legs and feet. Apply the gauze and use vet wrap to hold in place. Check the dressing daily, as the chickens may peck at it or it may get dirty.
- **Clear ointment:** Antibacterial ointment like Neosporin (get the kind without the pain-relief element) are great for protecting abrasions from infection.
- **Rice sock:** A sock filled with rice and tied off can be warmed in the microwave and used to warm a cold bird. You can also put it under the bottom of an eggbound hen while she is resting. The heat helps open the vent, which may help the egg pass.
- **Wax-based salve:** Coating combs and waddles with a wax-based salve provides a layer that repels frost and helps prevent frostbite. Buy a product like Nanak's Skin Repair from the store or make your own following recipes available online.
- **Disposable gloves:** Always wear gloves when treating your chickens' injuries, as some issues are caused by bacteria that can be transmitted to humans.
- **Blue antiseptic spray:** Antiseptic sprays with a blue staining agent help protect wounds from bacteria as well as camouflage red sores to help

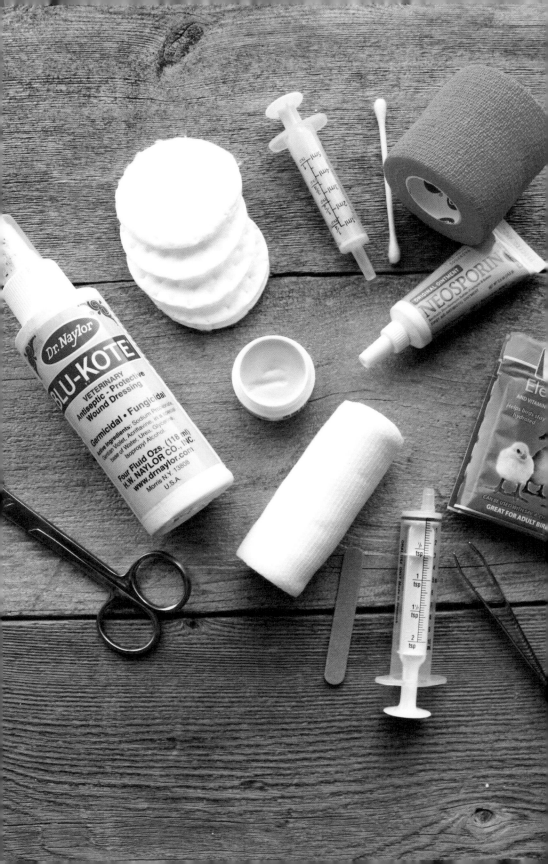

prevent pecking. You can buy this at the feed store or make it using recipes online.

- **Tweezers:** For removing debris and serving as a tiny set of hands when needed.
- **Needleless syringes and droppers:** Perfect for irrigating wounds, dosing medication, feeding ill chickens a liquid diet, and more.
- **Scissors:** For cutting bandages and gauze. Even though I own a dozen pairs of scissors, I like having a pair in the first-aid kit so I don't have to scramble to find them while dealing with an injured bird.

68
How to Set Up a Chicken Sick Bay

If one of your chickens falls ill or gets injured, it is in the best interest of both the individual chicken and the flock to separate them. Just like the first-aid kit, it's a good idea to have all the supplies for a sick bay ready to go before you actually need them so you aren't left scrambling when something happens.

What your sick bay needs
Your chicken sick bay should have the following things:

- A box, cage, or other way to contain the chicken
- Enough space for the chicken to stand up and move around
- Comfortable bedding
- Food
- Water

A large dog crate, rabbit pen, feed trough, or large plastic box all make good sick bays. If you are using a trough or a box, your sick bay should also have a lid that allows airflow. Lids are necessary to make sure the chicken stays put to focus on recovering. A piece of chicken wire or hardware cloth shaped to fit over the top makes a great lid.

In addition to making sure you have all these supplies on hand, it's also a good idea to know where you will set up sick bay in the event it is needed. This will keep you from scrambling when a chicken emergency does arise.

Sick bay should be set some place with low lighting and warmer than it is outside. Fill it with fresh bedding for each occupant. You will need to refresh the bedding daily and be sure to empty and disinfect it once your chicken no longer needs to use it.

69
How to Get Flies Under Control

Chickens bring all kinds of things into your life when they move in. Eggs. Joy. Silly antics. And flies. Poop attracts flies, and since chickens are adorable little pooping machines, your fly population will increase by approximately a gazillion the first summer you have chickens living outside.

A great fly-control strategy involves both repelling and trapping flies. No fly repellent works 100% of the time, so the trap is great for nabbing the ones that didn't get the hint from the repellent.

Fly repellent
You can make your own natural fly repellent by mixing the following together in a 16-ounce spray bottle:

- 1 cup apple cider vinegar
- 15 drops lavender essential oil
- 10 drops peppermint essential oil
- 10 drops basil essential oil
- 10 drops rosemary essential oil
- Fill to just below the top with water and add 1 Tablespoon dish soap

Shake the mix together before you use it. Spray it around your coop and run and on bedding (just a mist—don't soak things!). Unlike synthetic mixes, it won't last for a week or more, so you will need to re-spray every few days. I know I would prefer to re-spray more often rather than use something with harsh chemicals, though, so the payoff is worth it to me!

Fly traps
There are lots of different styles of fly traps on the market from sticky strips to electronic traps that draw flies in and then dehydrate them (which you can then feed to your chickens!). The trap I have personally found to be the best combination of affordable and effective is a bait-style trap called the Victor Fly Magnet. When I set it up in my yard, I was amazed at how many flies it trapped in a short period of time. It doesn't smell good, so it isn't suitable to use inside or in enclosed spaces, but it's perfect to use in an open-air run.

70
Egg-Eating Hen Solutions

If you started keeping chickens to have fresh eggs, it becomes incredibly annoying if the chickens beat you to the eggs first! Sometimes the egg eating is a symptom of a bigger problem, so use this as a chance to check and make sure you are meeting all their needs.

Why chickens eat their eggs

Once a chicken realizes how delicious eggs are, it's hard to get her to stop. She may have broken the egg by accident and happened to taste it. She may have pecked it out of boredom and realized it was a tasty snack. Overcrowding can also cause issues with egg eating.

How to stop egg eating

If you're currently dealing with an egg-eating hen, you can try a few things:

- Place ceramic or wooden eggs in the nest boxes. Pecking these isn't rewarding, so chickens learn that pecking eggs is boring.
- Collect eggs multiple times daily to keep fresh eggs from being broken and eaten.
- Build nest boxes that allow the egg to roll back and out of reach of chickens when laid.
- Hang curtains over nest boxes so hens aren't tempted to walk by nests and see them as snack buffets.
- Increase the amount of space your birds have access to, in case egg eating is due to overcrowding.
- Increase protein and/or calcium in their diet, in case egg eating is due to nutritional deficiency.
- Provide your birds with more activity and range access, in case egg eating is due to boredom.

If you have tried all these things and still have a bird eating eggs, you can either cull her or put her with chickens who aren't laying. Be warned that this is behavior that can be contagious and taught to the other birds. Be sure to figure out quickly which of your chickens is eating eggs before the behavior spreads.

71

Sanitation to Keep You and Your Flock Healthy

Both chickens and humans carry a lot of germs. To protect each other from getting sick, follow these simple preventative measures.

Wash hands

Always wash your hands before handling chicks or ill chickens. You should also wash your hands after handling chickens, doing chicken chores, or touching anything your chickens may have walked or pooped on.

Chicken shoes

We don't wear shoes in my house, but if you do, you should have a dedicated pair of shoes that you wear only for going outdoors around your chickens. Chicken poo can transmit salmonella, *E.coli*, and a host of other unpleasant diseases, and you definitely don't want to track that around your house. Keep extra shoes like clogs or boots on hand for guests for the same reason.

If you visit an area with other birds like a poultry show, a fair, or the house of your friend who has chickens, do not wear the shoes you wear around your birds. This helps protect the birds at the other location as well as your own flock once you return home. Once you return home, be sure to clean and disinfect the shoes you did wear to avoid introducing new pathogens into your chickens' environment.

Sanitize used items

If you purchase or borrow secondhand equipment that has previously been used around other people's birds, clean and sanitize it before it gets to your property. In the case of something lent to you, be sure to clean and sanitize it before giving it back to your friend.

72
Quarantine New Arrivals

With the exception of brand new chicks from a commercial hatchery or feed store,* any new arrivals to your flock must be quarantined for a minimum of thirty days. A quarantine period is important even if the new bird appears disease- and parasite-free. This will give you time to observe it on an ongoing basis and examine it multiple times. A quarantine period gives any diseases the chicken might be carrying (but not yet showing symptoms of) time to emerge.

While implementing the quarantine protocol is admittedly annoying, it is still essential. Introducing a sick or parasite-infested chicken to your flock could infect or wipe out your whole flock.

How to set up quarantine

A separate, safe, and predator-proof coop should be set up for the quarantined arrivals. The new coop should be located as far away from your established chickens as possible.

- Do not share equipment between the two coop areas.
- Thoroughly wash your hands and change your clothes after interacting with the chickens in the quarantine coop. Some people will also shower.
- Have a dedicated pair of shoes for using around the quarantined chickens that is different than the shoes you use to visit your established chickens. Make sure your paths to get to/from the established and quarantine chickens do not cross each other.
- Observe the new chickens for signs of illness or parasites.
- Once thirty days have passed, if your new chickens still appear healthy and disease-free after a thorough examination, take one or two "tribute" birds from your established flock and moved them in with or next to the quarantine birds in the quarantine area.
- Continue with your quarantine procedure for another two to four weeks. If all birds in the quarantine coop still appear healthy, you can work to move them in with the established flock.

* Biosecurity purists will still say that all birds, regardless of age, should be quarantined before introducing them to your established flock. I would find it very unusual for hatchery or feedstore chicks that are only a few days old to be contagious, which is why they are the one exception I personally make to a minimum thirty-day quarantine.

73
Essential Oils for Chickens

People aren't the only ones who can benefit from using essential oils. Certain essential oils can help chickens with a variety of common issues. This is a healthy and natural way to heal chickens and even clean their coops. Not all essential oils should be used around chickens but these are the most common and useful:

- **Lavender oil** helps heal and protect against sores and wounds on chickens. Using a little lavender oil diluted down to 1% can help the healing process of wounds and scratches.
- **Frankincense oil** can be a natural pain reliever. This oil must also be diluted to 1% or less. Frankincense oil should be used on closed wounds and in small amounts. A little oil will go far, especially on chickens.
- **Peppermint oil** can repel flies and unwanted insects. See page 145 for my recipe.
- **Tea tree oil** will repel unwanted mites in and around the coop. Dilute the oil with water, and just like the peppermint oil, spray around the coop.
- **Lemon oil** is a natural way to clean and disinfect your coop. No need for harsh chemicals when you have lemon oil! Dilute this oil down with water and spray inside the coop and around the run.

You should only use essential oils when they are properly diluted. For chickens, this means that the solution should be concentrated to 1% or less. A 1% solution can be made with 1 drop of essential oil and 1 teaspoon of a carrier oil such as coconut oil. When in doubt, always dilute more than you think might be needed.

Because chicken respiratory systems are highly sensitive, any essential oils used on the chickens must be kept away from eyes, mouths, and beaks. When spraying in or around a coop, the essential oil must be highly diluted with water and only sprayed when chickens are out of the coop. It's also a good idea to give the coop time to air out before the chickens return. Use these oils in small doses and see how beneficial they can be to your flock!

Gardening with Chickens

Chickens can be both a garden friend and a garden foe. Helpful in the spring for preparing a garden bed for planting, chickens can fluff dirt and eat weed seeds, pests, and more. Left alone too long near favorite plants, though, they will eat them to the ground. They might then dig up the roots just for good measure. Try the following tips to set yourself up for success, so gardening with chickens is more of a joy than a burden.

74
Grow an Herb Garden for Your Chickens

Herbs are some of my chickens' favorite plants. You can use herbs for all kinds of different things in chicken keeping. If they are growing in an area where the chickens can get to them, be sure to protect them so the chickens don't dig them up or eat every last part.

Things you can use herbs for
Chickens love it when you use herbs to keep them happy and healthy. Here are some of my chickens' favorite ways to use these powerful plants:

- Snacking straight off the bush
- In nesting boxes
- Brewed as tea
- Frozen into ice cubes for a summertime treat
- Sprinkled over their bedding
- Nibbling on leftover trimmings

Herbs chickens like
Basically, any herbs or flowers that humans can eat will be safe for chickens. These are some of my favorite herbs to grow for chickens:

- Basil
- Bee balm
- Calendula
- Catnip
- Chamomile
- Chives
- Dill
- Lavender
- Lemon balm
- Mint
- Oregano
- Parsley
- Rosemary
- Sage
- Thyme

75
Don't Get Burned by Poop

Chicken poop is a valuable gardening commodity. Its high nitrogen content is great for plants, but adding too much straight chicken manure can cause plant "burn." To get the most bang from your chicken-manure buck, try one of these composting techniques.

The sit-and-simmer pile
The deep-litter method (see page 53) only requires twice yearly clean-outs, so this is the primary composting system that I use. To use this method, simply clean out the coop litter and poop together into a large pile or bin. Allow the manure and litter to sit together and age for approximately six to twelve months before adding to your garden.

If you use aren't using deep litter, you can still use this approach with the addition of a dry carbon source such as dry leaves or grass clippings. Dump the chicken poop you clean out of your coop into a composting bin and layer it with dry leaves or grass clippings. If you stir the pile on a regular basis, it will be ready to use in five to six months. If you let it sit as-is, it will be ready to mix into your garden in nine to twelve months.

Composting in place
If you let your garden rest during the winter, composting in place is another great way to manage coop poop. In the fall, either till fresh manure and used coop bedding directly into the garden or bury it under a layer of topsoil. Allow the manure to compost in place in the garden all winter, and in the spring, the soil will be ready for planting.

Chicken manure tea
Making chicken manure tea is another way to process chicken poop to avoid burning plants. To make it, fill a burlap sack with chicken poo and add either large rocks or a brick to help it sink. Place the bag in a 20-gallon trash can that you fill with water and cover. Take the lid off daily to stir and aerate the mix. After three weeks, remove the burlap sack (you can move the poo to your compost system) and stir the mix well. Use mix as a liquid fertilizer on beds you are preparing for planting or plants that need a nitrogen boost.

76
How to Protect Plants from Chickens

On page 47, I talk about how hardware cloth is your hero because it does just about everything people think chicken wire should do. The one thing chicken wire is really good for, though, is protecting plants from chickens. Here are three different ways you can use chicken wire to protect your plants from your favorite tiny dinosaurs:

- **Cloche:** Traditionally made from glass to protect plants from dropping temperatures, cloches have been used in gardens since the 1800s. Make your own cloches out of chicken wire to protect freestanding plants from your chickens. I like to use these over herbs I have growing in areas my chickens have access to. The herbs can grow through the chicken wire, which lets the chickens munch any outside growth while still protecting the base of the plant.
- **Stem defense:** Plants that are growing against a wall or fence can benefit from having the lower 30 inches or so protected by a piece of chicken wire. Simply unroll a piece of chicken wire and bend it to make a half-cylinder shape. Secure the wire to the fence to help keep chickens from eating your vines and other plants to the ground.
- **Ground cover:** Chickens love scratching in the dirt and digging things up. For large plants, such as rose bushes or young trees, the chickens are unlikely to eat the plant itself, but too much scratching near the roots can kill the plant. Protect the roots of plants with vulnerable root systems by cutting a piece of chicken wire and laying it around the base of the plant. You can cover the wire with dirt for an improved appearance but the chickens won't be able to scratch below the wire and dig up the root system. You can also protect roots of vulnerable plants by placing large rocks or bricks around the base.

While each of these systems works for protecting individual plants, they aren't terribly practical for protecting entire gardens. If you have an entire garden that needs protecting, it is best to keep your chickens secured in an area away from it unless you want to use their help in the spring or fall to dig up the whole thing.

If you're unable to secure the chickens away from the garden, large row or bed covers can be fashioned from chicken wire or bird netting. If you try this method, be sure the covers are well anchored, as I have had chickens knock covers over or dig underneath them to gain access to the garden I was trying to protect.

77

Use Your Chickens to Prepare Garden Beds for Planting

Spring garden chores commonly include weeding and mixing in soil amendments in preparation for the upcoming growing season. Now that I have tiny weed-eating and soil-tilling machines, I have much less work to do myself!

How to put them to work

To get these benefits yourself, give chickens access to your garden beds for several days in the spring. If you normally give them a large range area, encourage them to stay in the garden by either creating a temporary barrier or by sprinkling treats on the ground in the garden. Flipping over a few sections of soil with a shovel often reveals bugs and worms and encourages the chickens to dig nearby to find more. After a day or two, every hornworm and weed seed should be cleared out by the chickens.

Next, if you plan to use natural soil amendments like peat moss, dried leaves, or compost, let your chickens spread it out and mix it into the soil for you. Chickens hate piles and will make short work of leveling any pile you put on your garden by kicking it around and digging through it looking for bugs. Don't use your chickens to spread chemical fertilizers, though. If you would need to wear gloves to prevent it from burning your hands, it isn't suitable for chickens to dig through.

Once your chickens have the soil prepared, you're ready for planting! Once you plant your starts of seeds, though, be sure to either restrict your chickens' access to the garden or cover areas where new plants are growing. Chickens love few things more than a fresh garden buffet and will quickly destroy fresh plants they are allowed access to.

78
Use Your Chickens to Clean Up Garden Beds

When the growing season is done and it's time to put your garden to bed for the winter, use your feathered friends to clean up. Just like they did in the spring, your chickens will love digging in your garden looking for bugs and weed seeds. While they work, they'll deposit fertilizer and till it into your soil.

How to put them to work

If you are using a chicken tractor–style coop, just park it on top of your spent garden beds. If you have a semi-free-range backyard flock like I do, just remove whatever barrier you were using to keep them out of your garden in the first place—open the gate or pull up the cover. If your chickens are normally cooped up or you live in a high predator area, pull up a chair (along with a book and a cool drink!) so you can sit outside and "chickysit" them. Your presence alone will usually be enough to keep predators from wanting to pick off one of your chickens as a snack. Some chicken owners build temporary enclosures that fit over garden beds to help contain and protect their chickens as they work in the garden from predators coming in by land or air.

Once your chickens have cleared away some of the spent vegetation, pile dead leaves onto your garden plot. Your chickens will have a blast digging through them and they will end up shredded and stirred into your soil as a result. This will give your soil a nice nitrogen boost, while the digging action of the chickens will help aerate it and break up clumps.

I also like to encourage my chickens to spend time working in my garden beds by feeding them there. I will sprinkle whole-grain feed or scratch grains over the garden so as they hunt for more food, they will encounter plants that need to be eaten back or bugs and worms that need to be dug up out of the ground.

Fresh Eggs

If you are reading this book, likely your end goal for owning chickens is to have baskets overflowing with fresh eggs. Beginning chicken-keepers often have lots of questions about keeping shells strong, storing eggs, and more. Even experienced chicken-keepers might learn a thing or two from these tips about how to keep their eggs fresh and organized!

79
Secret Ingredients for Strong Shells

As an egg is forming on its way through your chicken, one if its last stops is the shell gland. Here, the shell will grow around the outside of the egg and membrane. The shell is made almost entirely from calcium carbonate. If your hen doesn't have enough calcium in her system, she might lay very fragile eggs or even an egg with no shell at all!

Secret ingredient: Calcium

Once your chickens become mature (around twenty weeks of age or so), it's important to make sure they have adequate calcium. While "layer" formula chicken feed should have calcium in it, giving extra calcium as a free choice supplement helps chickens stay topped off as needed.

Ways to give chickens extra calcium can include:

- **Eggshells:** Save the shells of the eggs you use, allowing them to dry out. Crush them up into small crumbles and offer them back to the chickens in a free choice dish.
- **Oyster shells:** The easiest way to give oyster shells is to buy crushed oyster shells from the feed store. Offer in a free choice dish or sprinkle over the ground in their run.
- **Limestone:** Ground limestone is another popular way to give calcium to chickens. As with eggshells and oyster shells, set it out in a free choice dish or sprinkle it over the ground in the run.

Secret ingredient: Sunshine

To properly absorb calcium, chickens need access to sunshine. This is why the path of the sun is something you need to consider when you set up your coop and run. Sunshine doesn't have to be direct. The light that makes its way through to earth on a cloudy day still counts! To make sure they are getting enough sun, don't keep them in a dark or heavily shaded area all the time.

80
The Enemies of Clean Eggs

When chickens lay eggs, there is a clear coat that goes around the outside of the shell called the *bloom*. If the bloom stays intact, it does a great job protecting the eggs from microorganisms that might want to use it for a breeding ground (which then subsequently could make you ill). While you can wash your eggs, keeping eggs clean so they don't have to be washed until right before eating is ideal!

Poop

Even though a chicken only has one opening through which it both excretes waste and lays eggs, most poop that gets on eggs doesn't get there while it is being laid. Poopy nesting boxes is a top reason eggs end up with poop on them. To keep nesting boxes poop-free, don't let your chickens sleep in them. If I have a bird get particularly adamant about sleeping in a nesting box, I might block it off by putting a small plastic bin in the way to block her access to it (and then remove it in the morning when I let them out again). Broody birds won't poop in nesting boxes but they won't lay eggs either! Go to page 129 for more info on what to do with a broody hen.

Mud

Trust me when I say that I know about mud! We live in Oregon, where the rainy season stretches from October until May or June (sometimes even July!)—we have a lot of mud. Keeping the mud in your run to a minimum (see page 189) will go a long way to helping. Keeping a path between the door of the coop and the nesting boxes clear or filled with fresh shavings or a clean alternative bedding will also help clean off a bird's feet on her way to deposit an egg in the nesting box.

Broken eggs

If an egg breaks in your nesting box, it causes all sorts of problems! Not only can it trigger an egg-eating frenzy (see page 147 for help with that), but it makes a really big mess in both the bedding and all over any other eggs that were there. To prevent broken eggs, be sure your chickens have plenty of access to calcium to keep their eggs strong. Also be sure to line nesting boxes with a soft bedding like pine shavings or a commercial nesting box lining to provide soft padding for the eggs to land on. If you find a broken egg in your nest, be sure to clean it up right way and replace any soiled bedding with fresh.

81
How to Store Eggs

Stored properly, fresh eggs can last up to six months! If that sounds like a long time, keep in mind that the eggs you buy at the store are often a full month old by the time they get on grocery store shelves. Between giving grocery stores time to sell them and you time to eat them, the store-bought eggs are often months old by the time they become an omelet.

Storing at room temperature

As long as the bloom is intact, fresh eggs can sit out at room temperature for about a month. The bloom will still be intact as long as the shell doesn't have any apparent damage and you have not washed the egg (which washes off the bloom). One of the reasons I try and help my chickens keep their eggs clean is so I can store eggs on the counter. While you *can* store dirty eggs at room temp, they just aren't as nice to look at.

Storing in the fridge

I don't like putting poop in my fridge, so if I decide I am going to refrigerate eggs, I take a look at them and wash them first as needed. You can wash eggs by lightly buffing the dirt off with fine grit sand paper, wiping them with a damp towel, or by gently rubbing them under warm running water if it's really dirty. Soap and other cleaning agents aren't really necessary; water does the trick. Some people who have used soap to clean eggs complain that it made the eggs taste like soap, so that's another good reason to avoid it. Towel-dry the eggs as needed and put them in a carton or bowl to store in the fridge.

Do **not** clean your eggs by allowing them to sit in water or by running them under cold water. Cold water causes the contents of the egg to shrink, which can suck bacteria in. Allowing eggs to sit in water makes it easier for bacteria to penetrate the shell, which increases your chances of getting ill.

Keeping them in order

Even though eggs last quite a while, I do my best to eat them in the order the chickens laid them. There are lots of fresh-egg storage solutions on the market, but two popular ones are using an egg skelter (which allows the eggs to roll toward the bottom, leaving the older eggs ready to grab while the fresher eggs are near the top) and simply putting your eggs in a carton on the counter, moving the older eggs toward the front every day and loading fresh eggs in the back.

82
How to Tell if Eggs Are Still Good

Maybe you found a secret nesting spot and you're not sure how long the eggs have been there. Or maybe you forget to keep your cartons in order in the fridge and now you aren't sure how old the eggs are in that one carton tucked in the back.

Things to toss right away

There are some signs that an egg is definitely bad and you shouldn't even move on to the float-testing phase. If your eggs do any of these things, get rid of them. You may want to dispose of them in a sealed bag in your outdoor trashcan just to be safe. Any eggs exhibiting these symptoms are rotten and may even explode!

- Stink
- Have a crack
- Feel slimy
- Make a sloshing sound when you gently shake it
- Stick to your egg carton when you try to remove it

Candling

You can read more about the specific technique of candling on page 23. Candling is important especially if you have a rooster. It's a reliable way to make sure that none of the eggs were in the process of developing into chicks. You can also candle unfertilized eggs to get an idea of how large the air pocket is (eggs with large air pockets are older). Dark-shelled eggs are hard to candle, so don't be alarmed if you can't see much in those!

Crack test

Occasionally an egg will pass the tests above but still have something off about it. If you don't feel confident about the freshness of the eggs, you can also do the crack test—but wait until you are actually ready to use the egg before doing this one!

For this test, crack the egg into a small bowl. Give it a sniff and look for anything unusual about it. If nothing looks off, go ahead and eat it. If the egg looks or smells strange, though, it's better to play it safe and discard it.

Float testing

Another way to test egg freshness is to put the eggs in a bowl or glass of warm water. If the egg sinks to the bottom, it is very fresh. If it stands on end or hovers in the middle, it is less fresh but still okay to eat (these eggs make great hard-boiled eggs). Eggs that float on the surface should be discarded.

To protect against bacteria getting into the shells, don't let your eggs soak in your float test water. Test them and then dry them off right away before putting them in the fridge.

83
Who Is Laying Eggs?

You might be deciding which chickens should be eaten and replaced with younger hens. Or maybe you are just curious which chickens lay which eggs. Here are some tips to tell.

Signs a hen is laying
For this one, you will need to get up close and personal with the backside of your chickens. Hens who are currently producing eggs will have vents that are larger and more moist than non-producing hens. If you have lots of one breed of chicken, this will be an easier determination to make as you can compare the hens to each other. If you have lots of different breeds of chickens, deciding whose vent indicates they are laying might be harder to do because different breeds have different body types and features than others. When in doubt, move on to the food coloring method.

Food coloring method
For this method, you need a small amount of petroleum jelly and food coloring. Mix a dab of the jelly with a drop of food coloring. Head out to the coop early before the hens have had a chance to lay any eggs and apply the colored jelly to the vent of the chicken you would like to mark. Choose a different color for each chicken if you want to mark different chickens. If you mark Henrietta's bottom in blue dye and you find an egg in the nesting box with blue streaks, you know Henrietta is definitely laying eggs.

Write down which colors you marked which hens with. The food coloring should last a few days and you can always reapply if needed. Keep in mind that some breeds don't lay eggs every day even at peak production so don't get too anxious sharpening your axe.

84
What Can I Do with All These Eggs?

All winter you fussed about how your chickens are freeloaders, but now it's spring and you don't know what to do with all the eggs you have on your hands! Here are some ideas for things you can do with a glut of eggs.

Save them

Eggs can keep fresh for about six months if stored in the refrigerator (see page 169 for tips on that!). Save extra eggs laid during the summer to help keep you from having to do the winter walk of shame every chicken-keeper dreads—having to go to the grocery store to buy eggs!

You can also keep cartons of eggs with the dates written on them. After eggs are a month or two old, they make great hard-boiled eggs that are easier to peel.

Eat them

Whip up some high-egg volume recipes like omelets, frittata, quiche, or egg salad. Some make-ahead recipes like egg bites or eggs cooked in a round mold for breakfast sandwiches freeze well and make great grab-and-go breakfast options for busy families. Check out the Resources section starting on page 211 for some of my favorite high-egg recipes.

Give them away

If you have neighbors who are good natured about your feathered friends, it's nice to share some of the egg bounty with them. Give your cleanest, nicest eggs to your neighbors to help keep them happy about the chickens living adjacent to their property.

Another place you can often give eggs away is to food pantries. Call ahead and see if your local food pantry will take donations of fresh eggs from backyard hens.

Sell them

This one can be a bit tricky depending on where you live, but if you have a lot of chickens and a lot of eggs, it might be worth the research! Different states, counties, and towns have different rules regarding selling eggs from backyard chickens. Call your local agriculture extension office to get more information on what rules and regulations would apply to you if you wanted to sell any of the eggs laid by your hens.

85
Where Are My Eggs?

If you bought chickens to have fresh eggs but your girls aren't delivering, it can be frustrating! Try troubleshooting your low egg volume by asking yourself these questions:

Are they too young?
While some breeds of chicken will start to lay as early as sixteen weeks, others can take much longer. Some chickens won't be ready to lay eggs until they are closer to six months (or about twenty-six weeks). We have one chicken who didn't start laying eggs regularly until she was almost ten months old!

Do they have enough food?
Making eggs is a demanding process for a chicken. They need nutrient-rich food with plenty of protein and calcium for their bodies to do it well. If you feed them daily and they finish everything you set out, try increasing the amount of feed you give to see if that makes a difference.

Is she broody?
If you see the same chicken sitting in the nesting box every time you go out to collect eggs, you may want to change up your routine to see if it's just a coincidence of schedules or if she is indeed brooding. Other sure signs of brooding include assuming a position affectionately referred to as the "angry pancake," where the chicken has spread itself out to try and appear flatter. Broody hens will also growl or fuss at you when you collect eggs they are sitting on. See page 129 for more information!

Are they molting?
Growing fresh feathers is hard work and most chickens will stop laying entirely during their molts. Check out page 127 for more information on how to support your girls through a molt.

Are they getting enough light?
There is a direct connection between how much light a chicken gets and how many eggs it produces. In northern regions when the days get quite short during the winter, egg production can drop dramatically even if it isn't very cold. Providing extra light is one solution that you can read about on page 201.

86
How to Get Perfect-Peel Fresh Eggs

Most people prefer the taste of fresh eggs over the older eggs you buy at the local grocer. The one place where older eggs perform better, though, is when you want to serve them hard-boiled.

As an egg ages, the membrane starts to pull away from the shell. This makes older eggs easier to peel when prepared as hard-boiled. One way to get easy-peel hard-boiled eggs is to store eggs in your fridge for four to eight weeks to allow them to age before boiling.

My preferred method lets me use eggs of any age and they peel perfectly every time!

How to steam eggs in a pressure cooker
The eggs in the picture to the left were cooked following this method. Despite the eggs being less than a week old, the peels came off beautifully! To get this great result, I used an Instant Pot countertop pressure cooker. To get perfect-peel pressure-cooker eggs, follow this procedure:

1. Place in the bottom the steamer rack that came with your Instant Pot or similar pressure cooker.
2. Fill your pot with as many eggs as you would like to cook. (Remember to stay below the maximum fill line, if your cooker has one.)
3. Add 1 or more cups of water. (Check the manual for your specific cooker to see what the liquid minimum is for cooking; a large-capacity cooker may require a minimum of 2 cups of water.)
4. Lock the lid and close the vent. Set your pressure cooker to cook at high pressure for 5 minutes.
5. Once the 5-minute cooking timer has gone off, allow the cooker to cool for 5 minutes before rotating the vent to release any additional steam. While the cooker is venting, prepare an ice bath by filling a large bowl with ice cubes and then topping it with cold water.
6. Remove the lid when safe to do so. Remove the eggs from the cooker to the ice bath. Allow them to sit in the ice bath for 5 minutes before either peeling or drying to store in the fridge in their shells.

Under the Weather

No matter where you go, there you are—and the weather is always there with you! Sunny, cloudy, or stormy. Cold, hot, or wet. All kinds of weather conditions can cause issues for you and your flock. I cover some of the most common weather challenges faced by chicken-keepers in North America and how to come through with a smile and fresh eggs!

87
Hot Weather Treats

When the first heat wave of summer hits western Oregon, I have a few tricks up my sleeve to keep my chickens happy and cool. Here are thirteen things I feed my hens when it's hot.

Anything freezer burnt
A heat wave is a great time to clean out your freezer. Anything chicken-safe that is freezer burnt or that you know you and your family just won't eat can go to the chickens (see page 84 for more on chicken safe foods).

Watermelon
Watermelon is a great, hydrating hot-weather treat. If you find watermelon on sale, buy an extra one for your hens. Set out halves of fresh or frozen watermelon and watch the chickens peck it until only the green shell is left. We save our watermelon rinds for the chickens as well because they like to eat the white part that most humans leave behind.

Flavored water
Freeze a pan of water filled with your chickens' favorite fresh treats like herb clippings, fruit and veggie scraps, peas, corn, berries, and more. Set the pan out the next day. As the water melts, they will have cold water to drink and they will eat the tidbits as they become exposed by the melting ice.

Frozen fruits and veggies
During the summer, I keep my eyes peeled for frozen fruits and vegetables that go on sale and then overbuy and give the abundance to my chickens. Frozen fruits and veggies my chickens enjoy include:

- Beans
- Blackberries
- Blueberries
- Corns (kernels or on the cob)
- Cubed carrots
- Peas
- Raspberries
- Squash chunks
- Strawberries
- Succotash

Take a look around and be creative and you'll be able to keep your chickens cool and spoiled all summer.

88
Break the Wind

Wind steals heat and health from your birds. Thankfully, with the help of a windbreak, you can stop the wind and keep your birds from becoming deeply chilled. Windbreaks will help prevent wind, snowdrifts, and driving rain from hurting your chickens. Windbreaks also keep drafts from your coop so your birds can stay warm while sleeping.

Types of windbreaks
There are many types of windbreaks from tarps to trees to deflect and slow down cold winter winds.

- **Tarps:** Heavy-duty tarps can be used down the side of chain-link fencing or wrapped around the sides of a coop.
- **Thick plastic:** Less durable year-round than a tarp, thick plastic like bubble wrap or moving plastic can be used to wrap drafty coops in areas that experience windy or rainy seasons where only a temporary buffer from the wind is needed.
- **Hay or straw bales:** Hay and straw bales can be stacked in different configurations. If you make stacks in the middle of the run or range area and cover the run with a tarp, the straw will stay dry. This makes it available to use as nesting material or mud control as needed throughout the year.
- **Fences:** Solid fences make great windbreaks. If your area is windy, building your coop and run against a fence cuts the wind blowing in from one side.
- **Dense shrubs or plants:** Dense shrubbery like juniper, red-tipped photinia, arborvitae, or other types of evergreen plants make good natural windbreaks. Growing them around the edges of your run or throughout your range area creates areas where the chickens can duck out of the wind during the day.

89
Winterize Your Chickens

Winter can be harsh on all animals, including your chickens. Before the cold hits your area, you should be prepared to change the way you care for your chickens and their coops to help everyone make it safely through until spring. Even the breeds advertised as "cold hardy" will still need help and protection. Follow these tips to keep your chickens safe and warm:

- **Protect their combs:** Combs and waddles are the most prone to frostbite, and using a beeswax-based, chicken-safe balm can prevent frostbite on your chickens. Frostbite won't just make them uncomfortable but can cause infections as well. Using a beeswax-based balm such Nanak's or a homemade balm will prevent this and keep your chickens happy.

- **Stay dry:** Keep the run and the chicken coop dry and free of puddles and water. Placing a tarp over an uncovered run can help prevent water from coming in and making puddles. Keeping your chickens dry during cold weather is essential so that they do not develop frostbite or bacterial infections.

- **Give them space:** Make sure your chicken coop has enough space for all your birds to roost. Chickens keep warm by roosting together. If you notice that any of your birds are on the floor of the coop then you may not have enough roost space. Your chickens will want to stay warm overnight by roosting together.

- **Use wide roosts:** One way to prevent chickens' feet from frostbite is to have a roost that is at least 3 inches wide. The wider roost lets chickens tuck their whole foot under their feathers instead of having their curling toes (on a smaller roost) exposed to the cold air.

- **Make a path:** Scatter straw or hay on top of the snow so that your chickens will have a place to walk. Typically, chickens do not like to walk on snow, but if you lay out hay they will be more apt to walk across it, as it will provide some insulation between the cold, damp snow and their feet. You could also try shoveling a walkway for them.

- **Give bedtime snacks:** Feed scratch grains in the evening to help keep your chickens warm. Whole grains take longer to digest, which causes their bodies to do more work and keeps their metabolism running hot.

Watch for predators: Keep an eye out for tracks in the snow and signs that predators have been near your chicken coops. During the winter, animals such as foxes, skunks, and raccoons are desperate for food, which might cause them to be braver and more apt to attempt breaking into your chicken buffet, er, coop.

90
Get Your Mud under Control

I live in Oregon, where it rains nearly every day between late October and May. Melting snow, rain, and poo can turn your yard or chicken run into a mucky mess. Prolonged exposure to soggy ground can cause disease or conditions like bumblefoot. Puddles can become breeding grounds for bacteria which can cause a variety of other infections and even kill your chickens. Mud can also become a hazard for the people walking outside to care for the chickens. There are lots of reasons why getting your mud situation under control is essential. Try one of more of these tips to help keep your flock safe while waiting for the ground to dry out:

- **Pick a well-draining coop location:** Before placing a chicken coop, consider the area carefully. Is the area uphill? Downhill? Will water pool up around it and the run? Take care to place the coop in the driest possible area. Read more about coop location considerations on page 70.
- **Straw or hay:** Hay and straw are a great way to absorb excess water and provide traction. Leave big clumps of hay or straw in different parts of the run and your chickens will take care of spreading it out for you.
- **Build a bridge:** Use scrap wood to create platforms or walking bridges. This will help keep your birds off the ground while the water dries up.
- **Rocks, bricks, or cement blocks:** If you see that water runs mainly from one area when it rains, use rocks or cement blocks as a barrier to divert the water away from your chickens to a more appropriate location.
- **Leaves or pine shavings**: Take advantage of fallen leaves or wood shavings. Gather leaves from your property. Many neighbors are happy to let you blow leaves off their lawn in exchange for keeping the leaves. You can also ask about getting wood shavings from your local county clean-up crews. Many places will let you take the wood chips from ground up limbs cleaned up after a storm for free. Simply spread them out all over the mud.
- **Cover your run**: Use plastic or tarps over the top of enclosed but uncovered runs to keep water out. This can help prevent the mud from even forming.

- **Ask local chicken-keepers:** Don't be afraid to ask for advice at the feed store or in online forums for local chicken-keepers in your area. Many of them will have insights about local materials that are available in abundance and work well in your climate. In Oregon, for example, many people use empty hazelnut shells from local nut farms in their runs. Not every state grows 99 percent of the nation's hazelnut supply, though, so unless you live in Oregon, hazelnut shells might be hard to come by. Your area will have its own specialty materials.
- **Move if needed:** Sometimes an area is just doomed. If your coop is set up in an area prone to flooding and nothing seems to be working to control the mud, consider relocating your coop either permanently or temporarily while you build up the elevation of the ground you have it on.

91
Heat-Proof Your Chicken Yard

Heat is more dangerous to your birds than cold, but luckily it isn't complicated to guard against. Heat-proofing your run will make a big difference in whether your chickens bake or survive the summer. Chickens that aren't being stressed out by the heat are also more likely to keep laying eggs even through the summer—no air-conditioner needed!

- **Create shade:** Use creative landscaping (see page 65) as well as roofing, tarps, or shade sails to create shady spots for your chickens. Chickens need places to escape direct sun, especially in the hottest parts of the day.
- **Keep the ground cool:** If your chicken run is paved or covered in sand and has a tendency to overheat, be sure to provide additional bedding during the hottest parts of the year. This will provide a comfortable level of insulation between the ground and your chickens' feet.
- **Make it breezy:** While keeping coops draft-free in winter is important, during the summer, the ability to have a cross breeze both in the run during the day and in the coop at night will help keep chickens from overheating. If your area doesn't naturally have breezes, consider adding a fan during the warmest parts of the year to keep air circulating nicely.
- **Keep water cool:** Keep their water dish or bucket in a shady spot to prevent their water from getting too warm. I use an insulated two-gallon water dispenser fitted with a chicken-watering attachment (you can see it on page 89). This helps keep the water cool in the summer and above freezing in the winter.

These tips are all great for managing heat year-round in warm climates or through the summer in seasonal ones. For more tips on surviving sudden onset heat, visit page 199.

92
Rainproof Your Coop

Waterproofing your coop means a nicer coop for both you and your birds. Luckily, it doesn't require tons of skill, effort, or finances to do, and it'll more than pay off in healthier birds. Keeping your coop dry means your investment will last longer. You won't have to deal with rot or mold issues, which can make you and your birds sick.

Rainproofing tips and tricks

- **Look for the light**: Before rainy season hits, head out to your coop on a sunny day. Look up at the roof and at the walls. If you can see light through any cracks, those are places that water can get in!
- **Look for leaks**: Even though it's convenient to stay inside when it's raining, make sure you gear up and head out to look in your coop at least once per season when it is raining. Take a look for any places that water is getting in the coop so you can fix things that are broken or seal cracks with caulk.
- **Caulk it up:** Even if your coop started out rainproof, as weather warps the materials your coop is made from, you may need to seal up cracks. Use outdoor-safe silicone caulk to seal any cracks you saw when you went looking for light and leaks.
- **Raise it up:** Installing your coop on top of a waterproof foundation will help your coop last longer. My coop is made of wood, so I built up a brick outline for a foundation so that the wood isn't ever sitting on soggy ground.
- **Deep bedding:** I love deep litter for lots of reasons, but one of them is how forgiving it is. If your coop does spring a small leak or some rain blows in through the windows, the litter can absorb it and help control the moisture while you fix the source of the problem. If a leak does happen, add lots of dry litter and toss scratch grains around the inside of the coop to encourage your chickens to stir and fluff the litter to help evenly distribute the moisture.
- **Keep ventilation:** While having a waterproof roof and walls is important, don't seal your coop up so tight that there isn't ventilation. Chickens release a lot of moisture as they breathe. It's important to make sure that moisture isn't going to condense and drip onto them.

It's always better to prevent leaks before they start. Even after assembling a brand new coop, be sure to give it the light check so that you can seal tiny cracks with caulk before they become big problems down the road. Stay on top of seasonal light and leak checks, and with proper attention and maintenance, your coop should last a long time.

93
How to Keep Water from Freezing

Water is essential to your birds' survival. Regardless of temperature, chickens need access to water for drinking. As a general rule of thumb, anytime your chickens have access to food, they need water. Lack of water causes chickens to lay fewer eggs. Water is also needed for proper digestion and to avoid an impacted crop. It can be tough to keep your chickens watered during the winter because any time the temperature dips below 32°F (0°C) you end up with a big ice block! Here are some wintertime hacks to try with your own flocks to keep the water flowing all winter.

- **Try an insulated waterer:** I use a two-gallon insulated water dispenser jug fitted with a chicken watering attachment (you can see a picture of it on page 89). The insulated container means it will take longer for the water to freeze. This is great in areas with mild winters where the weather doesn't stay below freezing for many days in a row.
- **Water bowl:** Use a bigger bowl or container for water during the winter. Bigger bowls will take longer to freeze than a small or shallow dish. Rubber or metal dishes are good for this, as they are less likely to crack due to changes in temperature.
- **Use the sun**: If possible, place your water container in direct sunlight. Sunny spots are warmer and may help prevent the water from freezing too fast.
- **Ping-pong balls**: Purchase an inexpensive pack of ping-pong balls from your local dollar store and place them in your chickens' water bowl. The ping-pong balls will float and move around on the water. Obviously, this will not help if you experience a hard freeze or several days of below-freezing temps in a row. It is, however, a good way to keep water from freezing as easily when daytime temperatures are above freezing but may dip down overnight.
- **Electricity**: If you are able to run electricity to your chicken coop, you can try using an electric livestock water heater to keep your water from freezing.
- **Start with warm water first**: If you are only facing a temporary cold snap, you could muddle through inexpensively by simply refilling the water dish every day (or multiple times daily if needed). Start by filling the dish with warm water so it takes longer to get down to freezing.

94
How to Survive a Heat Wave

If warm weather isn't already a normal thing for your birds, a heat wave can come as a shock to their systems. If long, hot summers (or a consistently warm climate) is something you deal with in your area, be sure to also check out the tips on page 193 for long-term heat strategies. For short-term heat waves, be sure to check out these ways to keep your flock from overheating.

- **Water:** The best thing you can do for your chickens on a hot day is provide ample amounts of water. Add ice cubes if you are able, but having warm water is better than having no water.
- **Ice:** Large blocks will last longer than handfuls of ice cubes. To make blocks of ice, fill empty milk jugs with water and freeze them overnight. You can also try freezing treats like chopped fruit, veggies, and herbs into blocks of ice. Set out the "treat-sicle" in a large dish at the beginning of the day. As the ice melts, the birds will have cold water to drink and cool snacks to peck at.
- **Shade:** If your chicken run landscaping doesn't naturally provide for shade, be sure to create shade for them with tarps or pop-up sun shades as needed.
- **Wet sand:** If you live in an area that tends to be hot, providing a baby pool full of sand that has been wet down and placed in the shade can provide some cool comfort. The chickens will scratch and dust bathe. Turning the sand over will bring the cooler sand at the bottom to the surface.
- **Sprinklers**: Just like children, chickens love playing in sprinklers on hot days. Set up a sprinkler or mister stand to help bring some cool relief to your hot birds.
- **Fans:** Keep the air moving to alleviate feelings of being overheated. Pick up an inexpensive fan at a thrift store and plug it in with an extension cord if needed. You may find your girls jockeying to see who gets the privilege of being closest to it. Just keep in mind that water and electricity do not mix! Be safe and use any water-based heat solutions separately from the fan.
- **Cold snacks:** I use heat waves as an opportunity to clean out my freezer! The food won't go to waste and the chickens get a nice snack to help them cool off—it's a win for everyone! Check out 183 for more ideas about what to feed your girls during a heat wave.

95
Lights in the Winter

Few things are as controversial in the realm of backyard chicken keeping as deciding to setting up lights for your chickens during the winter. Some people use lights to help extend the "daylight" hours a chicken is exposed to to help it continue to lay eggs throughout the winter. Others say that chickens should be allowed to take the winter off from laying eggs so they can rest up in preparation for the spring.

Cautions with lights

If you do decide to use lights in your chicken coop, be sure to follow these precautions to keep your chickens happy, healthy, and safe.

- **Don't use lights that emit heat:** Use Christmas lights or LED lights to reduce your risk of fire from a hot bulb.
- **Turn off the lights during their molt**: Your chickens won't be laying eggs during their molting period no matter how much light they get. While your chickens are molting, turn off the lights to reduce stress and help them use their energy re-growing feathers.
- **Put the lights on a timer:** It isn't natural for the sun to be up all time. Make sure to give your chickens time to rest when the light isn't on. Set your lights on a timer so they automatically turn themselves off at a certain time every day. Aim to give your chickens no more than about sixteen hours of light a day—so if the sun is up for twelve hours, use the lights for a maximum of four extra hours a night.
- **Work up to it:** Don't just decide to start leaving lights on for an extra five hours every night. Start with just thirty to sixty extra minutes of light at first. Work up in thirty- to sixty-minute increments until you are using the amount of extra lighting you were aiming for.

Cock-a-Doodle-Do – Raising Roosters (or Not)

It is a common myth that you need to keep a rooster around if you want your hens to lay eggs. The truth is that roosters are completely optional members of your flock. In many areas, keeping roosters is not permitted. If you live in an area where you could be allowed to own a rooster if you would like, consider some of these tidbits when deciding if keeping a rooster is right for you.

96
Do You Need a Rooster?

Have you had a friend tell you that you need a rooster or your chickens will never lay eggs? Maybe you have even believed this common myth yourself. While some chicken-keepers enjoy having roosters, they definitely are not a thing you *need* unless your goal is to have fertilized eggs for hatching chicks. Here are some pros and cons to consider when deciding if you want to keep a rooster as part of your flock.

Rooster Pros

- Fertilizes eggs, which is necessary for hatching chicks
- Helps look out for predators
- Can defend flock in case of predator attack
- Nice to look at

Rooster Cons

- May not be permitted in your area
- Noise—crowing happens as early as 2 a.m. and all throughout the day
- Can injure hens during mating
- Can injure humans—children are especially susceptible

If you do decide to keep a rooster, be sure to keep a ratio of about 1 rooster to 8 to 15 hens. This helps keep individual hens from being mated too aggressively, increasing their chances of injury. If you have only ten hens and one rooster, and you end up with an extra rooster by accident, you will either need to remove one of them from the flock or get more hens (chicken math, anyone?). See page 209 for ideas on how to get rid of surprise roosters.

97
Playing Hot Potato with a Rooster

No, not the game where you toss something back and forth (surely no rooster would tolerate such a thing!). If your rooster's spurs are getting very long, it may be time to de-spur your rooster. Long spurs can injure people and hens alike. If long spurs have grown out at an unusual angle, it can also cause problems for the rooster when he walks.

While there are lots of ways to remove rooster spurs, many people swear by this method, which uses the heat and moisture from a hot potato to help them gently twist off without hurting the rooster.

What you need
Tub with warm water

- 1–2 potatoes
- Microwave
- Pot mitt or towel
- Pliers
- Diatomaceous earth (DE)

How to do it

1. **Get your potato ready** by microwaving it for about 8 minutes (give or take a few minutes depending on potato size) and then allow it to rest for another 2 minutes. This will give you some time to catch your rooster.
2. **Catch your rooster.** If your rooster is not trained to come when you call, it may be easiest to pluck him off the roost after he has settled in for the night.
3. **Wash his feet and legs** in the tub of warm water to get dirt and debris off his legs.
4. **Using an oven mitt or towel, grab the potato** and slide it onto one of your rooster's spurs. Push it on as far down the spur you can go without actually touching the potato to his leg.
5. **Hold the potato on his spur** for about 5 minutes. With the potato still on the spur, gently twist the potato to see if the spur has any give. If not, continue to hold the potato on for another 2 minutes and try again. Usually

by about 12 minutes, the potato has softened the spur enough to twist off with the potato.

6. **If 15 minutes has gone by** and the spur hasn't twisted off with the potato yet, remove the potato and use pliers at the base of the spur to gently twist it off. You will see the soft quick underneath—don't touch it, as it will be very sensitive. While the quick will be red and "angry" looking, you shouldn't see more than a drop or two of blood (if any at all). If the quick starts bleeding, sprinkle DE over the quick to help encourage clotting.

7. **Repeat for the other spur.** You may be able to re-heat the first potato and use it for the second spur. If the potato is too soft or too mangled from being used on the first spur, it's good to have a second potato as a backup!

98
What to Do with Surprise Roosters

Everyone loves surprises! Unless, of course, the surprise is that a chicken you were hoping would be a hen is actually a roo. If a rooster was not in your chicken-keeping plans, think about these options:

Keep him
If the laws and regulations in your area don't bar you from keeping a rooster, you could consider keeping him if you have at least eight hens in your flock. Check out page 204 to read about some of the pros and cons of rooster ownership.

Rooster collar
A variety of rooster collars are available on the market that constrict your rooster's throat just enough so that he cannot crow and he becomes about as loud as a hen. Many roosters do not care for these collars, though, and the battle with your rooster about wearing the collar may not be worth the effort. An improperly fitted collar does have the potential to really hurt your rooster, too, so be sure if you want to try this that you follow all instructions. Check your rooster frequently during the first week or so that he is wearing it to make sure he is breathing okay.

Eat him
If you wouldn't mind eating him, you could keep him for a while to fatten him up before eating him. How to harvest chickens is beyond the scope of this book, but there are many online demonstrations that should prove helpful if you decide you want to try butchering him yourself. If butchering isn't something you think you can stomach, reach out to other chicken-keepers via Facebook. Someone might be willing to either teach you or to take care of it for you for pay.

Rehome him
Even if a rooster isn't a good fit for you, it doesn't mean that someone else wouldn't welcome him into their flock. You can try making an ad for him on your favorite online resource for buying from your neighbors, like Craigslist or Nextdoor. Some feed stores also keep a list of people who have expressed interest in taking accidental roosters, so check with your local feed stores to see if they can help.

99
Sending a Rooster to Freezer Camp

Many people find roosters to be a wonderful addition to their flock. They protect and care for the hens and will often sacrifice themselves to save a hen. They're also fun to watch, as they will call out to their harem when they find choice treats.

Even if you are a happy rooster owner, there may come a time when your rooster needs to make his way to *Freezer Camp*—a family-friendly term often used to refer to butchering and processing your rooster.

Signs it's time to send him to Freezer Camp:

- **He didn't overwinter well:** If he suffered badly at the hands of winter weather, process him before fall.
- **You have too many roos:** Each rooster does well with a harem of eight to fifteen hens. If you do not have enough hens for each rooster, decide who you want to father your next round of chicks and process the others. Keep one juvenile backup rooster to replace your lead rooster in the future.
- **Aggression:** If your rooster tries to attack you or your child, he needs to go. Aggressive roosters can blind or permanently disfigure a person. Because they are shorter, children are especially vulnerable to being injured by aggressive roosters.
- **He's the wrong breed:** If you end up with a surprise rooster that isn't a breed you want, it's okay to harvest him.
- **He's eating more than working:** If your rooster isn't actively protecting hens or sharing with them, you may want to get rid of him to make room for a more considerate rooster.

Once you have identified the birds you'd like to get rid of, feed them well until they are at least twelve weeks old. There is no wrong or right time to process a chicken after they are about three months old. Each stage and age of growth provides benefits and a unique flavor. Younger birds are better for roasting, while older birds make good soup. The final question to ask yourself is if you would like to process him yourself or pay someone else to do it for you.

Eggs-tra Resources

I'll be the first to admit that I am not the only person who knows about raising chickens! There are many wonderful resources available in print and on the Internet to help you dig deeper into some of the topics in this book. I've also included a chicken wishlist worksheet and recipes for an overabundance of eggs. Enjoy!

100
Chicken People Are the Best People

It's really true! There's something magical about owning chickens that ties the owners together. Whether you live in the city or on a farm, chicken people are almost always willing to offer time and advice and share solutions that work for them.

Connecting with a larger community of chicken people is easy to do on the Internet. Here are some of my favorite online spots for finding others who understand the magic of poultry:

- **Instagram:** Who doesn't love digging through pictures of adorable chickens? You can connect with my chickens on Instagram at @CreativeGreenChickens. Poke through some hashtags like #chickensofinstagram and you'll find plenty of chicken people to follow in no time.
- **Facebook:** Facebook groups are wonderful places to connect with other chicken owners. The Internet changes so rapidly, I won't name my favorite chicken groups here, but if you go to Facebook and do a search for "chickens," lots of different group options will pop up! Try a couple and see how you like them, as each group has its own personality and rules of engagement.
- **Blogs:** There is a wealth of knowledge to be found on chicken blogs and websites. Independent bloggers are my favorite, but some brands like My Pet Chicken have really helpful blogs with lots of great information for beginners. Some of my favorite blogs for chicken content are:

  - *Creative Green Living*: This is my blog! In addition to beginner-friendly chicken-keeping content, I write about gardening, recycled crafts, and healthy options for families.
  - *Fresh Eggs Daily*: This blog is run by Lisa Steele, the author of multiple books about chickens. Lisa's blog shares beautiful pictures of her flock and natural chicken-keeping solutions that really work.
  - *Tilly's Nest*: Melissa Caughy writes about keeping chickens, gardening, and her work as an author of multiple chicken books.
  - *My Pet Chicken*: My Pet Chicken sells both live chickens and chicken-keeping supplies online. I find their blog easy to navigate with lots of beginner-friendly content.

101
My Favorite Chicken Books

There are so many wonderful books about chickens currently on the market! As of the writing of this book in 2018, here is a short list of my favorite books about keeping chickens. Pick up a copy from your local library or bookseller.

- *A Kids Guide to Keeping Chickens* by **Melissa Caughy (Storey, 2015).** Designed for children, this book is written in an approachable way that both kids and adults can understand. Melissa includes lots of helpful information and recipes as well. This is the Chicken King's favorite chicken book!
- *Let's Hatch Chicks* by **Lisa Steele (Quarto, 2018).** Beautifully illustrated by Perry Taylor, Lisa and her chicken, Violet, teach readers about hatching eggs with hens or in an incubator in an easy-to-understand way.
- *Backyard Chickens: Beyond the Basics* by **Pam Freeman (Quarto, 2017).** An editor and contributor to *Backyard Poultry Magazine,* Pam explains about keeping chickens in enough detail to keep you informed but not so much detail you get overwhelmed.
- *Storey's Guide to Raising Chickens* by **Gail Damerow (Storey, 2017 – 4th edition).** Weighing in at more than 400 pages of dense information, this book is a comprehensive and useful reference to all things chicken. From raising babies to first-aid, problem solving, butchering, and more, this book will have the answers to any question you have about chickens.
- *Gardening with Chickens* by **Lisa Steele (Voyageur, 2016).** Lisa is a master gardener, aspiring herbalist, and queen of all things chicken. This book covers everything you ever wanted to know about how to get chickens and gardens to coexist in a symbiotic relationship.
- *50 Do-It-Yourself Projects for Keeping Chickens* by **Janet Garman (Skyhorse, 2018).** Janet's book is an invaluable resource for the DIY-minded chicken keeper. She teaches you how to make your own chicken-keeping staples like feeders, waterers, and first-aid balm but also covers some fun projects like chicken swings and salad bars.

My Chicken Wishlist

Use this wishlist to make notes about chickens you read about in this book and online so you can keep them in mind while browsing for new breeds to build your flock!

Breed	Eggs/year	Egg color

Notes

_____

_____

_____

_____

_____

_____

_____

_____

Egg Recipes

Here are some of my favorite recipes that use lots of eggs!

Omelet Muffins

These muffins are a great grab-and-go breakfast. They also freeze beautifully and can be reheated when you would like to eat them. As written, the recipe makes twelve muffins and can be easily doubled to make more.

Ingredients

- 9 eggs
- 8 ounces cooked sausage or bacon, diced or crumbled
- 1 small red bell pepper, diced
- ½ red onion, diced
- Salt and pepper to taste
- ¾ cup shredded cheese

Directions

1. Preheat oven to 350°F.
2. Whisk eggs in a medium bowl. Mix in diced meat and vegetables. Season with salt and pepper to taste.
3. Line a muffin tin with baking cups and spray with olive oil.
4. Fill cups with egg mixture to just below the rim and top with approximately 1 tablespoon shredded cheese per muffin.
5. Bake at 350°F for 18 to 25 minutes.
6. Serve hot or allow to cool completely before removing from baking cups and placing in a large ziplock bag to store in the freezer.

Oven Baked French Toast

This recipe can be made for same-day consumption or be prepared the night before. It yields eight servings, so if you are expecting lots of guests, make two pans to use up a full dozen eggs.

Ingredients

- 1 loaf French Bread
- 1 tsp vanilla extract
- 6 eggs
- 1 cup milk

Directions

1. Preheat oven to 350°F.
2. Dice the French bread into large cubes and place in a large, well-greased casserole dish.
3. In a medium bowl, whisk vanilla, eggs, and milk until well combined.
4. Pour the egg mixture over the bread.
5. Allow to sit at least 15 minutes before baking so the bread can absorb the egg. You could also cover in tinfoil and refrigerate it overnight for easy baking in the morning.
6. Bake for 40 to 50 minutes until the egg is set and edges are golden brown.
7. Serve with maple syrup or fresh fruit.

Breakfast Casserole

This filling breakfast casserole is three breakfast favorites—hashbrowns, eggs, and bacon—rolled into one! Yield: 12 servings

Ingredients

- 2 Tbsp avocado oil
- 1 package (about 22 ounces) frozen toaster hashbrowns or tater tots, thawed
- 2 cups grated cheddar or Colby-jack cheese, divided
- 8 ounces cream cheese, softened
- 12 eggs
- Salt and pepper to taste
- 12 slices bacon, cooked and rough chopped
- 5 green onions, sliced, divided
- 3 roma tomatoes, seeded and diced

Directions

1. Preheat oven to 450°F. Grease a large casserole dish or sheet pan with the oil.
2. Crumble thawed hashbrowns or tater tots over casserole dish and press into the bottom of the pan. Top with half the shredded cheese.
3. Bake for 14 to 16 minutes or until crust begins to brown.
4. While crust is baking, mix cream cheese in a large bowl until smooth. Add eggs, salt, and pepper and whisk until well combined. Mix in bacon and all but 1/3 cup of green onion slices.
5. Once crust has been removed from oven, pour egg mix over crust and bake for an additional 8 to 12 minutes or until center is completely set.
6. Remove casserole from oven and top with diced tomatoes and remaining shredded cheese and green onion slices.

Deviled Eggs

This classic party appetizer is perfect anytime you are looking for a high protein snack! This recipe yields 24 egg cups.

Ingredients

- 12 eggs, hard-boiled (see instructions on page 179)
- ½ cup mayonnaise
- 1 Tbsp Dijon mustard
- Salt to taste
- Paprika or diced scallions for garnish

Directions

1. Peel the hard-boiled eggs and cut in half lengthwise.
2. Remove yolks to a medium bowl and mash with a fork or the back of a spoon. Mix in mayonnaise, mustard, and salt. Continue mixing until very creamy and consistent throughout.
3. Spoon yolk mixture back into the cavities in the cooked egg whites. For a more elegant look, use a frosting bag and star tip to dispense the yolk.
4. Garnish with dashes of paprika or diced scallions.

Acknowledgments

Nobody writes a book by themselves and this book is no exception. I'm grateful to my husband and kids, who patiently let me put off most of my normal responsibilities for several weeks at the end of the manuscript writing process. I was "almost done" writing this book for much longer than I anticipated and they graciously gave me the space I needed to finish.

I'm beyond thankful for the team that loved on my kids so I could have marathon writing sessions uninterrupted by the responsibilities of motherhood. My dad and stepmom, Russ and Michelle Berg, and my friend, Sally Taber, watched my kids on several different weekends and did all kinds of things with them that were much more fun than watching Mom type words into a computer. I really could not have finished writing this book without their help.

I had several writing assistants who lent their expertise, helped me get my thoughts organized, did some research and fact-checking, and had input on ways to improve the book. Each of these women is an experienced chicken-keeper in her own right and this book is better for their contribution to this process: Melissa Potvin, Ericka Potvin, Honey Rowland, and Tamara Rubin—thank you!

My friends Erin and Ryan Bradley, as well as Suzanne Franklin of Sunny Side Up Chicken Emporium, were wonderfully helpful in teaching me a lot about chicks when we got started with our first batch of babies several years ago. They were also incredibly patient and kind with teaching my oldest son about chicks. I'm not sure my husband is glad he agreed to let me have chicks for my birthday that year, but I definitely am.

Finally, thank you to everyone who was enthusiastic and encouraging about seeing me write a second book. My editor, Nicole Frail, and the team at Skyhorse were delightful to work with a second time. The Creative Green Living Tribe members, blog readers, and fellow green bloggers who shared encouraging words and promised to buy this book really made a difference when I was knee-deep in a book that turned out to be nearly three times longer than I had expected! I hope I didn't let you down.

Index

protein, sources of, 127
pullets, 3, 15, 49
pumpkin, 93
Purina, 91

Q
quarantine period, 151

R
raccoon-resistant coop
    locks, 136
rainproofing, 194
raspberries, 183
recipes, 217–220
    breakfast casserole,
        219
    deviled eggs, 220
    French toast, oven
        baked, 218
    omelet muffins, 217
red star breed, *see*
    golden comet
red-tipped phontinia,
    63
reflective ribbon, 137
refrigerator, storing
    eggs in the, 169
remote-control motor,
    51
Rhode Island red, 5, 7,
    12, 104
rice sock, 139
rock band, chicken, 113
room temperature, stor-
    ing eggs at, 169
rooster collars, 209
rooster eggs, 120
roosters
    aggression in, 37
    butchering and pro-
        cessing, 210
    pros and cons of hav-
        ing, 204
    ratio of hens to, 204

removing spurs
    from, 205–207
    what to do with, 209
roosts/roosting
    in the chicken run,
        123
    in the incorrect place,
        tricks for, 67
    as requirement in
        chicken coop, 45
    width, 50
    winter protection
        and, 187
rosemary essential oil,
    145
rosemary, 89, 154
rose petals, 59
Roundup®, 77
run
    bird netting over, 61
    hardware cloth over,
        61
    mud and, 189
    preventing mud from
        forming on, 189
    rye berries, 87

S
safety shelters, 63
sage, 59, 154
salmon favorolles, 5
salpingitis, 121
salve, 139, 187
sand, 53, 57, 105, 193,
    199
sanitation precautions,
    149
scarecrows, 137–138
scare tape, 137
scattering feed, 83
scissors, in first-aid kit,
    141
secret nests, 119
shade, 193, 199

sheds, turned into
    chicken coops, 45
shell-less eggs, 121
shrubs, 63, 185
sick bay, 143
slide-out boards, 57
snack block, 83, 91–92
snacks
    bedtime, 50
    hot weather, 183
    keeping chick-
        ens warm with
        bedtime, 187
    turned into an
        activity, 83
snacktivity, 83, 91, 123
soldier fly larvae, 107,
    127
spearmint, 59
sprinklers, 199
spurs, removing from
    roosters, 205–207
squash chunks, 183
steaming, 179
Steele, Lisa, 21, 81,
    213
*Storey's Guide to
    Raising Chickens*
    (Damerow), 215
straw, 53, 123, 187,
    189
straw bales, 185
strawberries, 183
stuffed animals, 35
Styrofoam, 84
succotash, 183
sunlight
    coop location and,
        70
    placing water in
        direct, 196
    strong eggshells and,
        165
supplements, 99

About the Author

Carissa Bonham lives near Portland, Oregon, with a house full of boys and a yard full of chickens. She runs the popular green and nontoxic lifestyle blog, *Creative Green Living*, where she teaches families how to make healthier choices that are beautiful, delicious, and really work.

Carissa and her husband sold their home in an HOA-controlled neighborhood to move to a house more conducive to chicken keeping in 2015. Her boys, Kaypha and Asher, have been enthusiastic participants in their family's chicken-keeping adventures. Carissa still has members of the original flock she and Kaypha raised as babies, including Queen Elsa, the white leghorn who appears on the cover. Carissa's chickens have their own Instagram account, @CreativeGreenChickens, where they post photos and sassy chicken commentary.